FIREFIGHTER FATALITIES IN THE UNITED STATES IN 2005

U.S. Department of Homeland Security

U.S. Fire Administration

August 2006

In memory of all firefighters
who answered their last call in 2005

To their families and friends

To their service and sacrifice

U.S. Fire Administration
Mission Statement

As an entity of the Department of Homeland Security, the mission of the USFA is to reduce life and economic losses due to fire and related emergencies, through leadership, advocacy, coordination, and support.

We serve the Nation independently, in coordination with other Federal agencies, and in partnership with fire protection and emergency service communities. With a commitment to excellence, we provide public education, training, technology and data initiatives.

ACKNOWLEDGMENTS

This study of firefighter fatalities would not have been possible without the cooperation and assistance of many members of the fire service across the United States. Members of individual fire departments, chief fire officers, U.S. Forest Service personnel, the Department of Justice, the National Fire Protection Association (NFPA), the National Fallen Firefighters Foundation, and many others contributed important information for this report.

C^2 Technologies, Inc., of Vienna, Virginia, conducted this analysis for the U.S. Fire Administration (USFA) under contract EME-2003-CO-0282.

The ultimate objective of this effort is to reduce the number of firefighter deaths through an increased awareness and understanding of the causes of the deaths, and how they can be prevented. Firefighting, rescue, and other types of emergency operations are essential activities in an inherently dangerous profession, and tragedies do occur. This is the risk all firefighters accept every time they respond to an emergency incident. However, the risk can be greatly reduced through efforts to improve training, emergency scene operations, and firefighter health and safety.

PHOTOGRAPHIC ACKNOWLEDGMENTS

The USFA would like to extend its thanks to Patti Odbert, Scientific and Commercial Systems Corporation, who provided many of the photographs in this report taken at the National Fallen Firefighters Memorial weekend in Emmitsburg, MD, October 2005.

TABLE OF CONTENTS

continued on next page

BACKGROUND

For 29 years, the U.S. Fire Administration (USFA) has tracked the number of firefighter fatalities and conducted an annual analysis. By collecting information on the causes of firefighter deaths, the USFA is able to focus on specific problems and to direct efforts toward finding solutions to reduce the number of firefighter fatalities in the future. This information also is used to measure the effectiveness of current programs directed toward firefighter health and safety.

One of the USFA's main program goals is a 25-percent reduction in firefighter fatalities in 5 years, and a 50-percent reduction within 10 years. The emphasis placed on these goals by the USFA is underscored by the fact that these goals represent one of the five major objectives that guide the actions of the USFA.

In addition to the analysis, the USFA provides a list of firefighter fatalities to the National Fallen Firefighters Foundation. If Memorial criteria are met, the fallen firefighter's next of kin, as well as members of the individual fire department involved, are invited to the annual Fallen Firefighters Memorial Service. The service is held at the National Emergency Training Center (NETC) in Emmitsburg, Maryland, during Fire Prevention Week. Additional information regarding the Memorial Service can be found by visiting www.firehero.org or by calling the National Fallen Firefighters Foundation at (301) 447-1365.

Other resources and information regarding firefighter fatalities, including current fatality notices, the National Fallen Firefighters Memorial database, and links to the Public Safety Officers' Benefits (PSOB) program can be found at http://www.usfa.dhs.gov/fatalities/

INTRODUCTION

This report continues a series of annual studies by the USFA of onduty firefighter fatalities in the United States.

The specific objective of this study is to identify all onduty firefighter fatalities that occurred in the United States and its protectorates in 2005, and to analyze the circumstances surrounding each occurrence. The study is intended to help identify approaches that could reduce the number of firefighter deaths in future years.

In addition to the 2005 overall findings, this study includes information on firefighter accountability programs.

WHO IS A FIREFIGHTER?

For the purposes of this study, the term "firefighter" covers all members of organized fire departments in all States, the District of Columbia, the Commonwealth of Puerto Rico, the Virgin Islands, American Samoa, the Commonwealth of the Northern Mariana Islands, and Guam. It includes career and volunteer firefighters; full-time public safety officers acting as firefighters; State, territory, and Federal Government fire service personnel, including wildland firefighters; and privately employed firefighters, including employees of contract fire departments and trained members of industrial fire brigades, whether full- or part-time. It also includes contract personnel working as firefighters or assigned to work in direct support of fire service organizations.

Under this definition, the study includes not only local and municipal firefighters, but also seasonal and full-time employees of the U.S. Forest Service, the Bureau of Land Management, the Bureau of Indian Affairs, the Bureau of Fish and Wildlife, the National Park Service, and State wildland agencies. The definition also includes prison inmates serving on firefighting crews; firefighters employed by other governmental agencies, such as the U.S. Department of Energy; military personnel performing assigned fire suppression activities; and civilian firefighters working at military installations.

WHAT CONSTITUTES AN ONDUTY FATALITY?

Onduty fatalities include any injury or illness sustained while on duty that proves fatal. The term "onduty" refers to being involved in operations at the scene of an emergency, whether it is a fire or nonfire incident; responding to or returning from an incident; performing other officially assigned duties such as training, maintenance, public education, inspection, investigations, court testimony, or fundraising; and being on call, under orders, or on standby duty, except at the individual's home or place of business. An individual who experiences a heart attack or other fatal injury at home as he or she prepares to respond to an emergency is considered on duty when the response begins. A firefighter who becomes ill while performing fire department duties and suffers a heart attack shortly after arriving home or at another location may be considered on duty, since the inception of the heart attack occurred while the firefighter was on duty.

On December 15, 2003, the President of the United States signed into law the Hometown Heroes Survivors Benefit Act of 2003. After being signed by the President, the Act became Public Law 108-182. The law presumes that a heart attack or stroke occurred in the line of duty if the firefighter was engaged in nonroutine stressful or strenuous physical activity while on duty and the firefighter becomes ill while on duty or within 24 hours after engaging in such activity. The full text of the law is available at: http://frwebgate.access.gpo.gov/cgi-bin/getdoc.cgi?dbname=108_cong_public_laws&docid=f:publ182.108.pdf

The inclusion criteria for this study will be affected by this change in the law. Previous to December 15, 2003, firefighters who became ill as a result of a heart attack or stroke after going off duty needed to register some complaint of not feeling well while still on duty in order to be included in this study. For firefighter fatalities after December 15, 2003, firefighters will be included in this study if they become ill as the result of a heart attack or stroke within 24 hours of a training activity or emergency response. Firefighters who become ill after going off duty in a situation in which their activities while on duty were limited to nonstressful tasks that did not involve physical exertion—such as tasks that were clerical, administrative, or nonmanual in nature—will not be included in this study.

A fatality may be caused directly by an accidental or intentional injury in either emergency or nonemergency circumstances, or it may be attributed to an occupationally related fatal illness. A common example of a fatal illness incurred on duty is a heart attack. Fatalities attributed to occupational illnesses also would include a communicable disease contracted while on duty

that proved fatal, when the disease could be attributed to a documented occupational exposure.

Injuries and illnesses are included even when death is considerably delayed after the original incident. When the incident and the death occur in different years, the analysis counts the fatality as having occurred in the year in which the incident took place.

Five firefighters died in 2005 as a result of incidents that had occurred in previous years. Information about these five deaths is included in the Appendix of this report, but they are not addressed in the body of the report unless the death has an impact on retrospective statistical comparisons.

There is no established mechanism for identifying fatalities that result from illnesses such as cancer that develop over long periods of time, which may be related to occupational exposure to hazardous materials or products of combustion. It has proved to be very difficult over the years to provide a complete evaluation of an occupational illness as a causal factor in firefighter deaths due to the following limitations: the exposure of firefighters to toxic hazards is not sufficiently tracked; the long-term effects of such toxic hazard exposures often are delayed; and firefighters also may receive such exposures while off duty.

SOURCES OF INITIAL NOTIFICATION

As an integral part of its ongoing program to collect and analyze fire data, USFA solicits information on firefighter fatalities directly from the fire service and from a wide range of other sources. These sources include the Public Safety Officers' Benefit (PSOB) program administered by the Department of Justice, the National Institute for Occupational Safety and Health (NIOSH), the Occupational Safety and Health Administration (OSHA), the U.S. military, the National Interagency Fire Center, and other Federal agencies.

The USFA receives notification of some deaths directly from fire departments, as well as from such fire service organizations as the International Association of Fire Chiefs (IAFC), the International Association of Fire Fighters (IAFF), the NFPA, the National Volunteer Fire Council (NVFC), State fire marshals, State training organizations, other State and local organizations, fire service Internet sites, news services, and fire service publications. The USFA also keeps track of fatal fire incidents as part of its Major Fires Investigation Program and performs an ongoing analysis of data from the National Fire Incident Reporting System (NFIRS).

PROCEDURE FOR INCLUDING A FATALITY IN THE STUDY

In most cases, after notification of a fatal incident, initial telephone contact is made with local authorities by the USFA to verify the incident, its location, jurisdiction, and the fire department or agency involved. Further information about the deceased firefighter and the incident may be obtained from the chief of the fire department or his or her designee over the phone, or by other data collection forms.

Information that is requested routinely includes NFIRS-1 (incident) and NFIRS-3 (fire service casualty) reports, the fire department's own incident reports and internal investigation reports, copies of death certificates or autopsy reports and results, special investigative reports, police reports, photographs and diagrams, and newspaper or media accounts of the incident. Information on the incident also may be gathered from NFPA or NIOSH reports.

After this information has been obtained, a determination is made as to whether the death qualifies as an onduty firefighter fatality according to the criteria previously described. With the exception of firefighter deaths after December 15, 2003, the criteria used for this study were the same as those used in previous annual studies. Additional information may be requested, either by follow up with the fire department directly, from State vital records offices, or from other agencies. The determination as to whether a fatality qualifies as an onduty death for inclusion in this statistical analysis is made by the USFA. The final determination as to whether a fatality qualifies as a line-of-duty death for inclusion in the Fallen Firefighters Memorial Service is made by the National Fallen Firefighters Foundation.

2005 FINDINGS

One hundred and fifteen (115) firefighters died while on duty in 2005. (Using the pre-Hometown Heroes criteria, 99 firefighters died while on duty in the United States.) This level of firefighter fatalities, while unacceptably high, represents a hopeful decline in the annual number of onduty firefighter fatalities, if the pre-Hometown Heroes fatality criteria are used.

In December of 2003, the Hometown Heroes Survivors Benefit Act of 2003 was signed into law. For Federal survivors' benefit purposes, the law presumes that a heart attack or stroke are in the line of duty if the firefighter was engaged in nonroutine stressful or strenuous physical activity while on duty and the firefighter becomes ill while on duty or within 24 hours of engaging in such activity. Prior to the Act, Federal survivors' benefits for firefighters were not generally paid for heart attacks or strokes, regardless of the circumstances.

The first firefighter to die in 2005 as a result of onduty activity was Captain Ornell Fuller of the Midway Volunteer Fire Department in Dexter, New Mexico. Captain Fuller died the day after a response to a structural fire.

The inclusion criteria for this report were also modified when the Act became law. Prior to December of 2003, a firefighter needed to express or show signs of illness prior to going off duty in order to be included in this report. After December of 2003, firefighters who became ill within 24 hours of onduty stressful or strenuous activity also were included.

While a change in report criteria does not diminish the sacrifices made by the firefighters who die, or the sacrifices made by their families and their peers, statistical analysis of death trends needs to acknowledge the change. If the Hometown Heroes criteria firefighter fatalities of 2005 are set aside for a moment, the number of firefighters who died while on duty in 2005 was 99. Using previous report inclusion criteria, this is the lowest level of onduty deaths since 1998. This is also the first time since 1998, using those previous criteria, that the number of onduty firefighter fatalities would be below 100. The lowest years on record are 1992, with 77 fatalities, and 1993, with 81 fatalities (Figure 1).

Five firefighters died in 2005 as a result of crime. Three firefighters died in arson-caused fires, one firefighter was fatally shot, and one firefighter was killed in a crash with a vehicle that was fleeing law enforcement.

The 115 deaths in 2005 resulted from a total of 109 incidents. There were four firefighter fatality incidents in which two or more firefighters were killed in 2005.

In 2001, 344 firefighters were killed as a result of the attacks on the World Trade Center (WTC) in New York City on September 11. When conducting

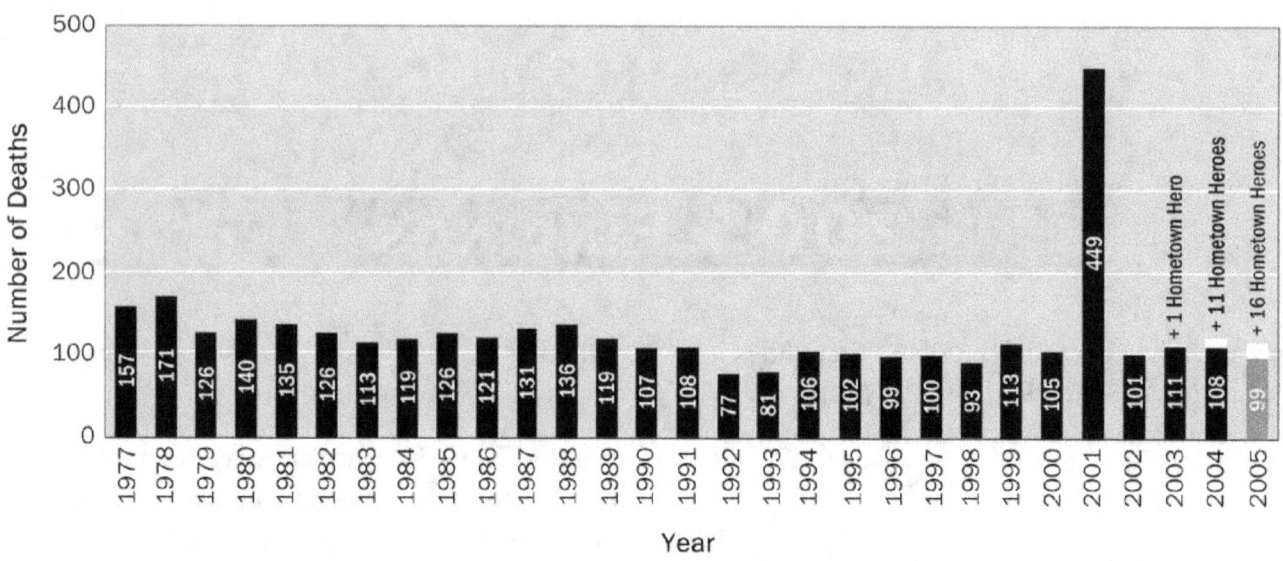

Figure 1. Onduty Firefighter Fatalities (1977-2005)

multiyear comparisons of firefighter fatalities in this report, it may be necessary to set these deaths apart for illustrative purposes. This action is by no means a minimization of the supreme sacrifice made by these firefighters.

CAREER AND VOLUNTEER DEATHS

Firefighter fatalities in 2005 include 81 volunteer firefighters and 34 career firefighters (Figure 2). Among the volunteer firefighter fatalities, 71 were from local or municipal volunteer fire departments, and 10 were members of wildland fire agencies. All of the career firefighters who died were members of local or municipal fire departments. Three of the firefighters who died in 2005 were female and 112 were male.

MULTIPLE FIREFIGHTER FATALITY INCIDENTS

The 115 deaths resulted from 109 incidents. There were four multiple firefighter fatality incidents in 2005, resulting in the deaths of 10 firefighters:

Table 1. Multiple Firefighter Fatality Incidents

Year	Number of Incidents	Total Number of Deaths
2005	4	10
2004	3	6
2003	7	20
2002	9	25
2001	8	362
2001 w/o WTC	7	18
2000	5	10
1999	6	22
1998	10	22
1997	8	17
1996	3	8
1995	7	18
1994	6	26

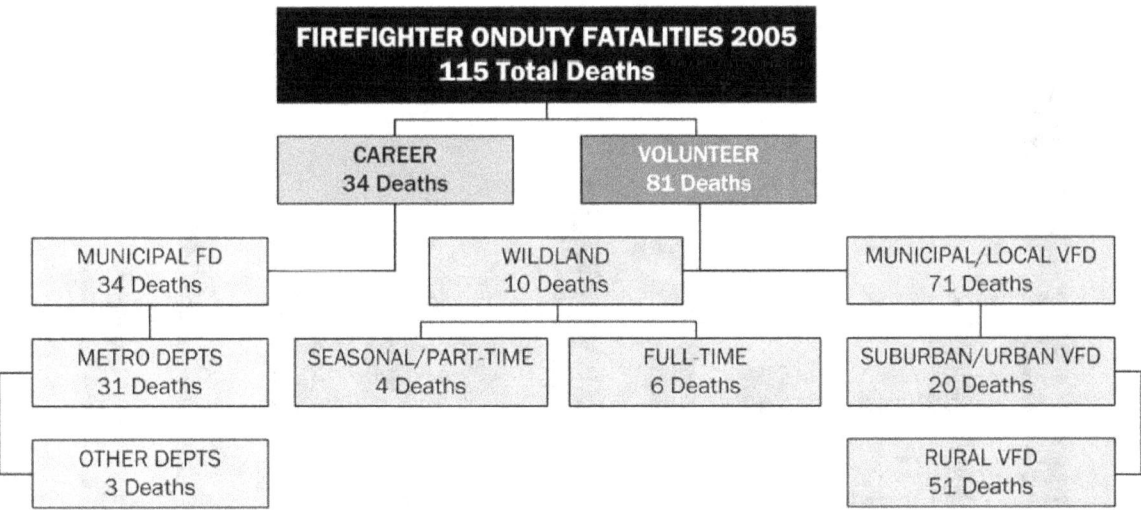

Figure 2. Career and Volunteer Firefighter Onduty Fatalities (2005)

- In January, two New York City firefighters died when they were trapped by rapid fire progress in a structure fire. Six firefighters were forced to jump from a fifth-story window because the apartment where they were trapped did not have a fire escape.

- In March, three firefighters who were employees of a private aircraft firefighting company were killed in the crash of their aircraft during a training exercise in California.

- In April, two Wyoming firefighters were killed when fire progressed rapidly in a townhouse fire.

- In April, three wildland firefighters were killed in the crash of a helicopter during a prescribed burn in Texas. The crew had been assigned to drop flammable spheres to initiate the burn when they experienced an unknown problem and crashed.

The New York City Fire Department suffered a third firefighter fatality on the same day as the January incident outlined above. The fatality occurred at a structure fire unrelated to the multiple-firefighter-fatality incident. The Memphis Fire Department lost two members in 2005 in separate incidents. The Memphis deaths were caused by a CVA (cerebrovascular accident, or stroke) and a heart attack.

WILDLAND FIREFIGHTING DEATHS

Nineteen firefighters died in 2005 while engaged in activities related to brush, grass, or wildland firefighting. This total includes part-time and seasonal wildland firefighters, full-time wildland firefighters, and municipal or volunteer firefighters whose deaths are related to a wildland fire (Figure 3).

After a year with no multiple-firefighter-fatality wildland incidents in 2004, six firefighters were killed in two aircraft crashes in 2005. One crash involved an airtanker and the other involved a helicopter. Both incidents involved activities in advance of a fire incident. The crash of the airtanker occurred during training and the helicopter crash occurred in support of a prescribed burn.

The number of fatal wildland incidents decreased from the levels seen in 2003 and 2004. There were 15 fatal wildland incidents in 2005, compared to 21 incidents in the previous 2 years:

- Six firefighters died of heart attacks related to wildland fire activities.
  - A Florida firefighter died after returning home from a work shift that included the overhaul of two wildland fires.

9

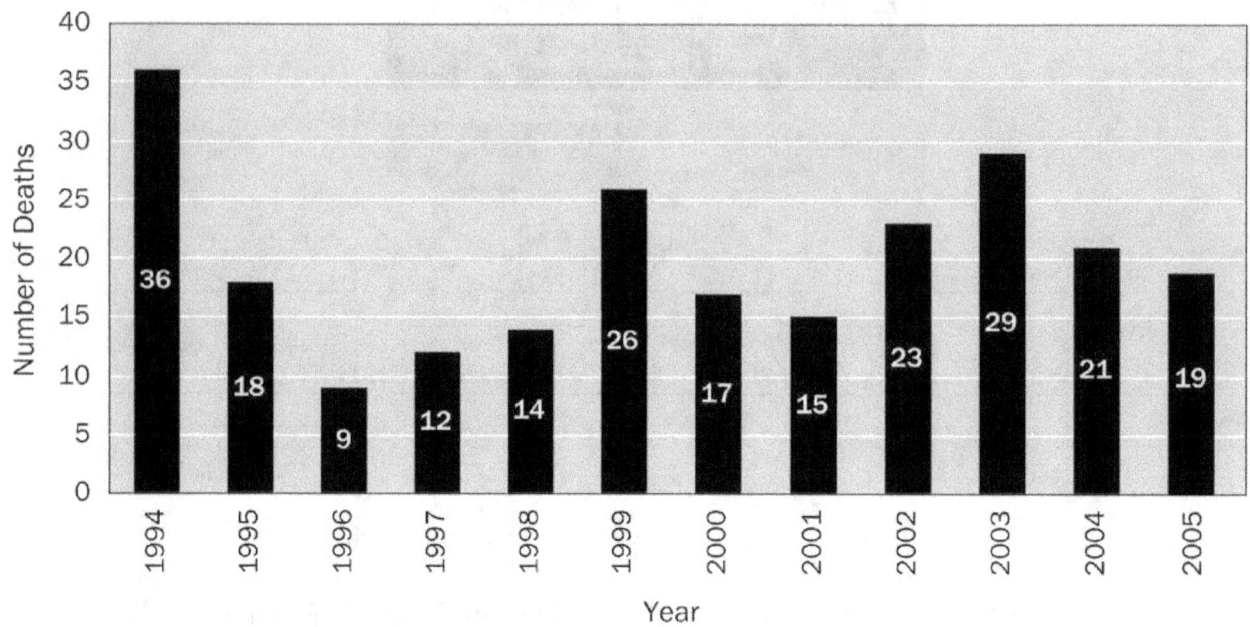

Figure 3. Firefighter Fatalities Related to Wildland Firefighting (1994–2005)
Note: Includes only firefighter fatalities related to fir

- A South Dakota firefighter was found dead in his hotel room after a work shift engaged as a Single Engine Air Tanker (SEAT) manager at a wildland fire in Colorado.
- A South Dakota firefighter collapsed of a heart attack after helping other firefighters fight a wildland fire on his land caused by a fire in a hay bailer.
- A New Mexico firefighter died of a heart attack during the final evaluations in a tree-felling class.
- An Oklahoma firefighter died after a solo fire fight at a wildland fire. Arriving firefighters witnessed his collapse.
- A Mississippi firefighter collapsed and died after completing extinguishment of a small wildland fire.
- Six firefighters died in wildland aircraft crashes.
  - Three firefighters were killed in the crash of an airtanker during training in California. The crew had completed a number of practice drops when the aircraft crashed for reasons unknown.
  - Three firefighters were killed in Texas after the crash of a helicopter. The aircraft was being operated in support of a prescribed burn.
- Three firefighters were killed in vehicle crashes related to wildland activities:
  - A Kansas firefighter was killed when the brush truck he was driving collided head-on with a fire department tanker (tender) responding to the same incident. The vision of both drivers was obscured by smoke.
  - A Texas firefighter was killed in the crash of a tractor-trailer water tanker while responding to a mutual-aid wildland fire.
  - A Nevada firefighter was killed when an all-terrain vehicle (ATV) that he was driving overturned as it was being driven downhill as the firefighter was supervising a wildland conservation project.

- A Kansas firefighter was killed after he called to report a wildland fire resulting from a lightning strike at his home. The firefighter went outside to investigate, contacted a live power line, and was fatally electrocuted.
- A Virginia firefighter was fatally burned as he fought a wildland fire. His body was discovered the next day, after he failed to return from his efforts.
- A Tennessee wildland firefighter was killed in a domestic violence incident as he waited for a piece of apparatus to be repaired. A gunman killed three people, including the firefighter, the gunman's wife, and another civilian.

Table 2. Firefighter Deaths Associated with Wildland Firefighting

Year	Total Number of Deaths	Number of Fatal Incidents	Number of Firefighters Killed in Multiple-Death Incidents
2005	19	15	6
2004	21	21	0
2003	29	21	10
2002	23	14	13
2001	15	9	9
2000	17	14	6
1999	26	25	2
1998	14	13	2
1997	12	10	4
1996	9	9	0
1995	18	14	7
1994	36	18	22

Table 3. Wildland Firefighting Aircraft Deaths

Year	Total Number of Deaths	Number of Fatal Incidents
2005	6	2
2004	3	3
2003	7	4
2002	6	3
2001	6	3
2000	6	5
1999	0	0
1998	3	2
1997	5	3
1996	0	0
1995	3	1
1994	8	3

TYPE OF DUTY

Activities related to emergency incidents resulted in the deaths of 60 firefighters in 2005 (Figure 5). This includes all firefighters who died while responding to an emergency, while at an emergency scene, while returning from an emergency incident, and in the course of other emergency-related activities. Nonemergency activities accounted for 55 fatalities. (Nonemergency duties include training, administrative activities, performing other functions that are not related to an emergency incident, and post-incident fatalities in which the firefighter had not experienced the illness during the emergency.) A multiyear historical perspective concerning the percentage of firefighter deaths that occurred during emergency duty is presented in Table 4.

11

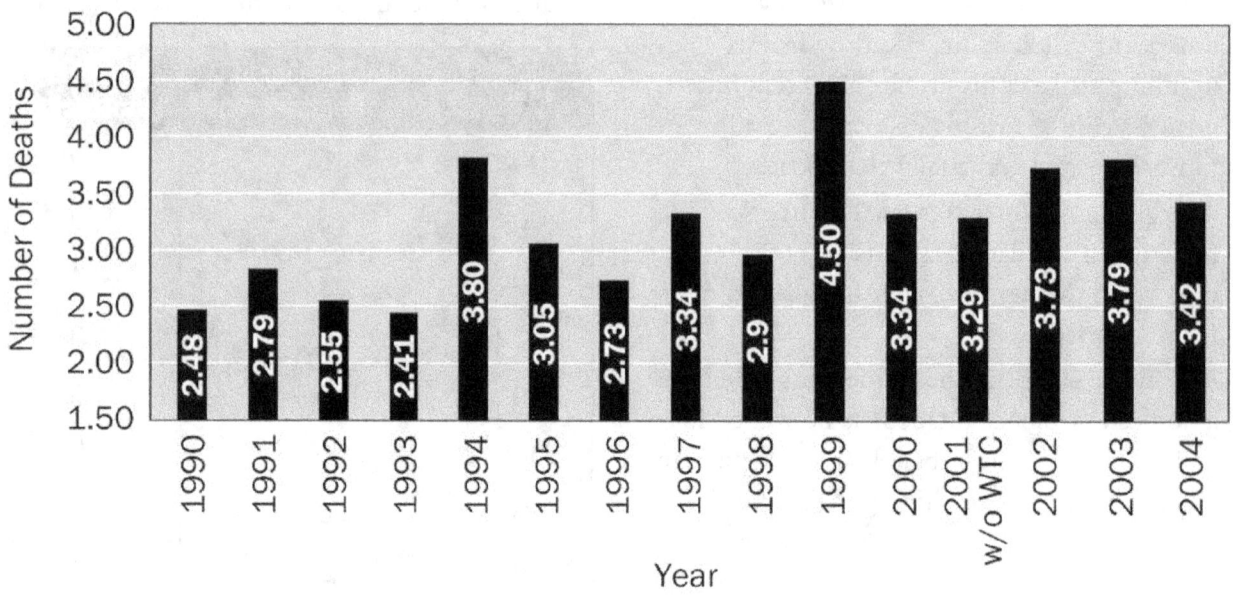

Figure 4. Firefighter Fatalities per 100,000 Fires

Table 4. Emergency Duty Firefighter Deaths

Year	Percentage of All Deaths
2005	52.2
2004	68.3
2003	70
2002	73
2001	65
2001 w/WTC	92
2000	71
1999	87
1998	77
1997	81
1996	72
1995	86
1994	84

The number of deaths by type of duty being performed in 2005 is shown in Table 5 and presented graphically in Figure 6. As has been the case for most years, fireground duties are the most common type of duty for firefighters killed while on duty.

If the Hometown Heroes deaths of 2005 are removed from the formula, the percentage of firefighter deaths associated with emergency duty rises to 60.6 percent.

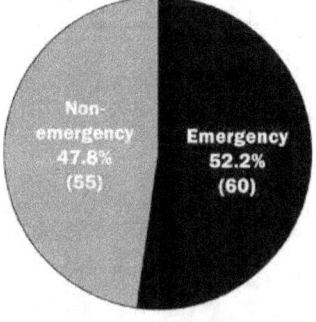

Figure 5. Firefighter Deaths by Type of Duty (2005)

Table 5. Firefighter Deaths by Type of Duty (2005)

Type of Duty	Number of Deaths
Fireground Operations	27
Responding/Returning	23
Other On Duty	24
Training	14
Nonfire Emergencies	6
After an Incident	21
Total	**115**

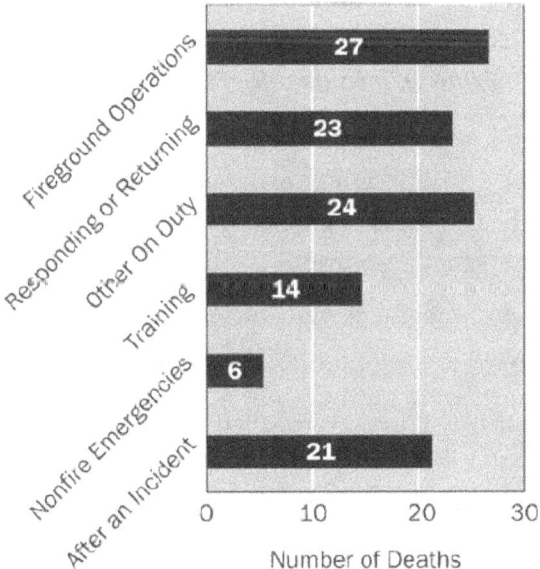

Figure 6. Fatalities by Type of Duty (2005)

Fireground Operations

Twenty-seven firefighters died while engaged in activities at the scene of a fire in 2005:

- Two New York City Fire Department firefighters were killed when they were trapped by fire progress in an occupied multiple dwelling. Firefighters were forced to make a five-story jump when their escape routes were cut off.

- Two Wyoming firefighters were killed when they were overcome by fire progress in a residential fire. The fire emerged from concealed spaces and extended rapidly.

- Thirteen firefighters suffered heart attacks at fire scenes in 2005:
  - Three of the heart attacks occurred at wildland fires.
  - Eight firefighters suffered heart attacks at fires in residential occupancies. Two of these fires had suspicious or arson-related causes.
  - A Delaware firefighter suffered a heart attack at an alarm activation incident.
  - An Arkansas firefighter suffered a heart attack at the scene of a car fire.

- Two firefighters were electrocuted at fire scenes in 2005:
  - A California firefighter was electrocuted when he came into contact with an energized wire at a residential structure fire.
  - A Kansas firefighter was killed after he called to report a wildland fire resulting from a lightning strike at his home. The firefighter went outside to investigate, contacted a live power line, and was fatally electrocuted.

- Two firefighters were killed when rapid changes in fire conditions trapped them. Both fires occurred in residential occupancies, one in New York and the other in Michigan.

- A Virginia firefighter was burned fatally as he fought a wildland fire. His body was discovered the next day, after he failed to return from his efforts.

- A Kentucky firefighter was killed when the fire apparatus he had driven rolled forward and crushed him at the scene of a residential structure fire.

- A North Carolina firefighter was killed when a fire-damaged tree limb crushed him as firefighters attempted to extinguish a fire in the tree.

- A Missouri firefighter became entangled in a man lift and was killed during a fire in a mill.
- A Texas firefighter was killed while advancing a hoseline in an abandoned residential structure. The roof of an addition collapsed under fire conditions and trapped the firefighter.
- A New York firefighter collapsed and died of a CVA that struck as he arrived on the scene of a working residential structure fire.

United States Firefighter Disorientation Study:

For the third year we include information regarding a report prepared by Captain William Mora of the San Antonio Fire Department who analyzed 17 incidents in which firefighters became disoriented in structures, resulting in a total of 23 firefighter fatalities. The study revealed a disorientation sequence common to all of these incidents that included light smoke showing upon arrival, an aggressive interior attack, deteriorating interior conditions, and firefighters becoming separated from handlines. The study recommends Standard Operating Procedures (SOP's) and training that includes a cautious initial assessment and a managed initial attack, if warranted. The full study is available at: http://www.sanantonio.gov/safd/pdf/FirefighterDisorientationStudy.pdf or through a link at the USFA firefighter fatality Web site, http://www.usfa.dhs.gov/fatalities/

Responding/Returning

Twenty-three firefighters died while responding to or returning from emergency incidents in 2005. Nineteen firefighters died while responding to emergency incidents, and four died while returning from emergencies. A comparison with previsous years is given in Table 6.

Four firefighters died while returning from incidents:

- A New York firefighter died when he had a heart attack and crashed the rescue truck that he was operating.
- A Kentucky firefighter experienced a heart attack and pulled to the side of the road in his personal vehicle while returning from an incident.
- A Tennessee firefighter experienced a heart attack while driving his engine company back to the fire station from an incident. The company officer was able to bring the apparatus to a safe stop and the crew provided treatment.

- A Georgia firefighter experienced a heart attack as he drove his engine company back from an emergency response.

Nineteen firefighters were killed while responding to incidents:

- Four firefighters were killed in crashes involving fire engines:
  - A Georgia firefighter going from one emergency to respond to another was killed when the right wheels of the engine went off the road and the apparatus crashed.

- A California firefighter was killed when his engine left the roadway and he was ejected from the rear passenger area of the apparatus.
- An Alabama firefighter was killed when the engine he was driving was involved in a crash with a tractor-trailer. The crash necessitated extensive extrication efforts to free the firefighter, who died of his injuries.
- A Louisiana firefighter was killed when his engine company was involved in a crash with a tractor-trailer. The firefighter was ejected from the officer's seat.

• Four firefighters were killed in crashes that involved fire department tankers (tenders) during response:
  - A Kansas firefighter was killed when the brush truck he was driving collided head-on with a fire department tanker (tender) responding to the same incident. The vision of both drivers was obscured by smoke.
  - A Texas firefighter was killed in the crash of a tractor-trailer water tanker while responding to a mutual-aid wildland fire.
  - A Mississippi firefighter was ejected from a tanker and killed when the tanker crashed during response.
  - A Texas firefighter was killed when the tanker he was driving blew a tire and crashed.

• Four firefighters became ill before they were able to leave their homes after being dispatched to emergencies. One of the deaths occurred in Maryland, one in Kentucky, and the other two in Pennsylvania. One was due to a pulmonary embolism and three were due to heart attacks.

• Three firefighters were killed in crashes involving their personal vehicles while responding to incidents:
  - A Pennsylvania firefighter was killed when his vehicle left the roadway, returned to

the roadway, and crashed into a command vehicle responding to the same incident from the opposite direction.
  - A Kentucky firefighter was killed when his personal vehicle left the right side of the roadway on a curve and crashed.
  - A California firefighter was killed when his personal vehicle left the roadway during response and crashed into a utility pole. The firefighter was ejected.

• Three firefighters experienced heart attacks while responding. In one case, the firefighter was driving his personal vehicle. In the other two cases, the firefighter was driving fire apparatus.

• A Texas firefighter was killed when he fell from a ladder truck as it turned during a response. The firefighter struck his head and died 2 days later.

Table 6. Firefighter Deaths While Responding To or Returning From an Incident

Year	Number of Firefighter Deaths
2005	23
2004	23
2003	36
2002	13
2001	23
2000	10
1999	26
1998	14
1997	21
1996	22
1995	29
1994	22

Other On Duty

A total of 24 firefighters died while engaged in other onduty activity in 2005. This includes deaths that were not associated with the response to any particular emergency.

- Three wildland firefighters were killed in the crash of a helicopter during a prescribed burn in Texas. The crew had been assigned to drop flammable spheres to initiate the burn when they experienced an unknown problem and crashed.
- Seven firefighters were killed in vehicle crashes while they were on duty but not assigned to an incident or training activity:
  - A Nevada firefighter was killed as he was supervising a wildland conservation project when an ATV that he was driving overturned as it was being driven downhill.
  - A Pennsylvania firefighter was killed as he returned from training when a truck crossed the center line of the roadway and crashed into his personal vehicle.
  - A Utah firefighter was killed when the tanker he was driving for maintenance experienced a tire failure and crashed.
  - A Missouri firefighter was killed during a fire department standby at a local raceway. A vehicle went out of control and the firefighter was crushed as he attempted to stop it.
  - An Alabama firefighter was killed when the rescue boat in which she was a passenger crashed into another boat as it returned from a standby at a boat parade.
  - A Missouri firefighter was killed when his vehicle was struck by another vehicle that was fleeing from law enforcement.
  - A New Hampshire firefighter was killed as he drove to work in response to an offduty shift recall when he lost control of his vehicle and crashed in poor weather.

- Seven firefighters suffered heart attacks while on duty but not working an incident. These heart attacks struck firefighters as they slept, as they set up for fire department functions, and as they participated in meetings and other fire department functions.
- Three firefighters suffered CVA's while on duty but not working an incident:
  - A Maryland firefighter suffered a CVA as he helped to unload supplies in preparation for a fire department event.
  - An Oregon firefighter suffered a CVA as he attended a fire department management conference.
  - A New Jersey firefighter suffered a CVA as he drove a pumper in the funeral procession of a fellow firefighter and friend who had died off duty.
- An Arizona firefighter died of an overdose of medications while on duty in the fire station.
- A Tennessee firefighter was shot and killed as he waited for an apparatus to be repaired.

16

- A Wisconsin firefighter suffered a pulmonary embolism and died during a town meeting as he represented the fire department at the meeting.
- A Delaware firefighter slipped and fell through a pole hole in the fire station while on duty. The firefighter later died of complications of the fall.

Training

Fourteen firefighters died while engaged in training in 2005. In most years, one or more firefighter deaths occur as firefighters complete a work capacity or "pack" test to qualify for wildland firefighting duties. No such deaths occurred in 2005.

- Three firefighters who were employees of a private aircraft firefighting company were killed in the crash of their aircraft during a training exercise in California.

- Six firefighters died of heart attacks related to training:
  - A New Jersey firefighter and a New Mexico firefighter died of heart attacks that struck during or immediately after physical fitness training.
  - A Florida firefighter was struck by a heart attack as he and his crew left a training facility after self-contained breathing apparatus (SCBA) training.
  - A Kentucky firefighter was struck by a heart attack after completing an SCBA training routine and before running other firefighters through the activity.
  - A Texas firefighter complained of not feeling well after training and later died.
  - A New Mexico firefighter suffered a heart attack and later died. The attack struck during final exercises in a tree-felling class.

- A Pennsylvania firefighter suffered severe burns and died in a simulated basement fire at a training facility.

- A Pennsylvania firefighter failed to surface during night dive rescue training and drowned.
- A Connecticut firefighter fell during training as he got on a ladder to descend to the ground, struck his head, and later died.
- A Florida firefighter suffered heat injuries during a physical fitness run while in recruit school; he later died of his injuries.
- A New York firefighter suffered a CVA as he participated in a Firefighter I class and later died.

Since 1990, some 19 firefighters have drowned on duty while engaged in water rescue operations. Seven of these drownings occurred during training. NIOSH has published an alert on dive rescue training. It is available at www.cdc.gov/ niosh/docs/wp-solutions/2004-152/

Table 7 offers a multiyear perspective on training deaths.

Table 7. Firefighter Deaths During Training

Year	Number of Firefighter Deaths
2005	14
2004	13
2003	12
2002	11
2001	14
2000	13
1999	3
1998	12
1997	5
1996	6
1995	3
1994	7

Nonfire Emergencies

Six firefighters died at nonfire emergencies:

Three firefighters died of heart attacks that struck them at nonfire emergencies:

- An Alabama firefighter collapsed as a result of a heart attack as he exited his personal vehicle at the scene of a motor vehicle crash.

- A South Carolina firefighter responded to an incident in which a body was discovered that had been dead for some time. The firefighter left the home where the body was found and collapsed of a heart attack.

- A New Jersey firefighter responded in his personal vehicle to a report of a gas leak. He evacuated residents from the home, exited the structure, and collapsed of a heart attack.

- An Iowa firefighter working on a farm was notified of a person who had collapsed in a manure pit. The firefighter activated the 9-1-1 system and entered the pit to attempt a rescue. The firefighter was overcome by manure gas; both men died.

- A Texas-based military firefighter responded to a rescue incident while deployed in Iraq. Firefighters were searching for soldiers trapped in an overturned vehicle in a canal; the firefighter drowned.

- A New Jersey fire police officer was struck and killed by a vehicle at the scene of a haz mat incident. The driver of the car was convicted of driving under the influence of alcohol and other offenses.

After the Incident

Twenty-one firefighters died after the conclusion of their onduty activities:

- Five of these deaths involved firefighters who complained of not feeling well, or exhibited signs of illness, prior to going off duty.

- Sixteen of these deaths did not involve any report of illness while on duty. These firefighter fatalities are being included in accordance with the Hometown Heroes inclusion criteria. This fact does not diminish the tragedy or importance of these deaths, but is pointed out for statistical analysis purposes only.

Of the 21 deaths of firefighters who died after the conclusion of duty, the vast majority were due to heart attacks.

- Eighteen of these firefighters died as a result of a heart attack.

- One firefighter died from a heart valve problem that may have been hereditary.

- One firefighter died of an overdose of prescribed pain medications.

- One firefighter died of an aortic aneurysm.

Career, Volunteer, and Wildland Deaths by Type of Duty

Figure 7 depicts career, volunteer, and wildland firefighter deaths by type of duty. Wildland career, wildland seasonal, and wildland contractor deaths were grouped together. As in past years, a disproportionate number of fatalities were experienced by volunteer firefighters responding to and returning from alarms, as compared to career firefighters. Fourteen volunteer firefighter deaths occurred while responding, and three occurred while returning from an emergency. Of the responding deaths, eight were due to vehicle crashes.

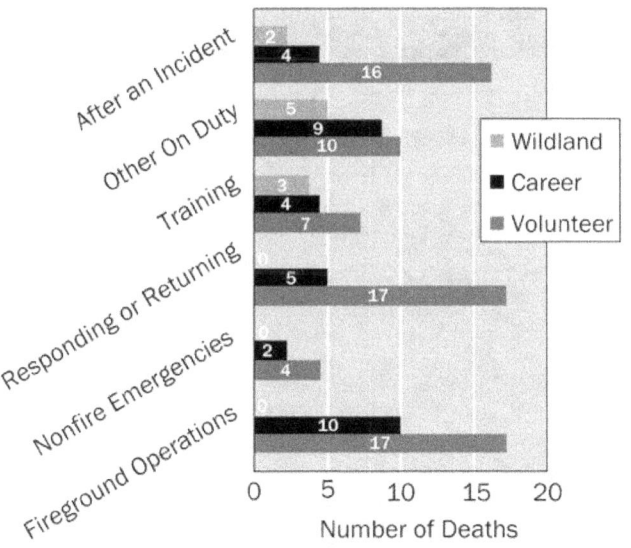

Figure 7. Career, Volunteer, and Wildland Deaths by Type of Duty (2005)

Type of Emergency Duty

In 2005, some 60 firefighters died while engaged directly in the delivery of emergency services. This number includes deaths that were the result of injuries sustained on the incident scene or en route to the incident scene, and deaths of firefighters who became ill on an incident scene and later died. It does not include firefighters who became ill or died while returning from an incident (such as a vehicle collision that occurred while the firefighter was returning from an incident). Figure 8 shows the number of firefighters killed in firefighting, emergency medical services (EMS), technical rescue-related incidents, and other emergency incidents in 2005. The type of emergency incident involved in one of the firefighter fatalities in 2005 is unknown.

Thirty-eight firefighters were killed in relation to fires; 15 firefighters were killed in relation to EMS calls; and 6 firefighters were killed at emergencies that involved hazardous materials and technical rescue, such as storm-related calls.

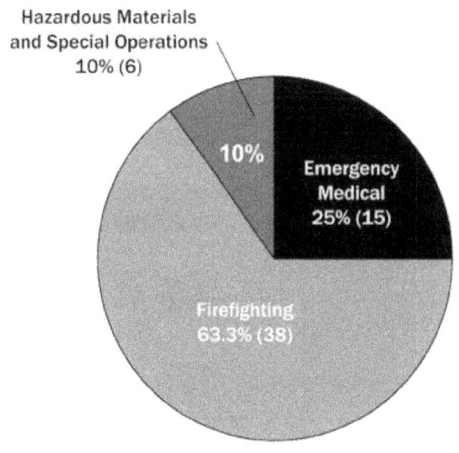

Figure 8. Type of Emergency Duty (2005)

Note: Some 60 of 115 deaths occurred during emergency responses, one incident type unknown.

CAUSE OF FATAL INJURY

The term "cause of injury" refers to the action, lack of action, or circumstances that resulted directly in the fatal injury. The term "nature of injury" refers to the medical cause of the fatal injury or illness, often referred to as the physiological cause of death. A fatal injury is usually the result of a chain of events, the first of which is recorded as the cause.

Table 8 and Figure 9 show the distribution of deaths by cause of fatal injury or illness.

Table 8. Cause of Fatal Injury (2005)

Cause	Number
Stress/Overexertion	62
Vehicle Collision	25
Caught/Trapped	9
Fall	5
Struck by	4
Contact/Exposure	3
Assault	1
Other	6
Total	**115**

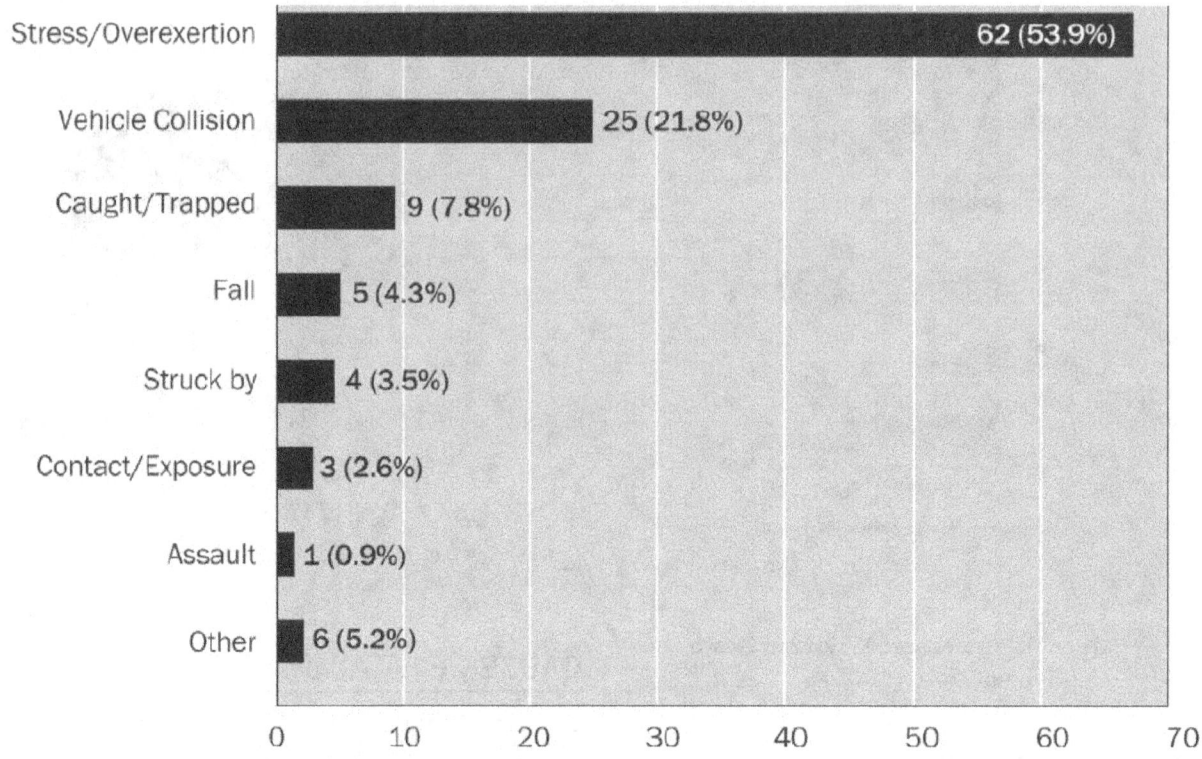

Figure 9. Fatalities by Cause of Fatal Injury (2005)

Stress or Overexertion

Stress or overexertion is a general category that includes all firefighter deaths that are cardiac or cerebrovascular in nature, such as heart attacks and strokes (CVA's), and other illnesses such as extreme climatic heat exposure. Classification of a firefighter fatality in this "cause of fatal injury" category does not indicate that a firefighter was in poor physical condition.

Firefighting is extremely strenuous physical work, and is probably one of the most physically demanding activities that the human body performs.

Firefighters who died on duty in 2005 of heart attacks and CVA's had a median of 21 years of service, and those who died of traumatic injuries had a median of 12 years of service.

- Sixty-two firefighters died in 2005 as a result of stress/overexertion:
  - Fifty-five of the stress deaths were heart attacks.
  - Six firefighter deaths were due to CVA's.
  - One firefighter fatality was due to heat exhaustion.
- If the Hometown Heroes deaths in 2005 are set aside for analysis purposes, 46.5 percent of firefighter fatalities in 2005 were caused by stress or overexertion (see Table 9).

Table 9. Deaths Caused by Stress or Overexertion

Year	Number	Percent of Fatalities
2005	62	53.9
2004	66	56.4
2003	51	45.9
2002	38	38
2001	43	40.9*
2000	46	44.6
1999	56	49.5
1998	43	46.2
1997	41	41
1996	46	46.4
1995	49	48
1994	36	34.2

* Does not include the firefighter deaths of September 11, 2001, in New York City.

Vehicle Crashes

As in most years, the second leading cause of fatal injury for firefighters who died in 2005 was vehicle crashes (see Figure 10):

- Twenty-five firefighters were killed in 2005 as a result of vehicle crashes.

- Seatbelts were used only in 8 of the 13 cases where the status of the firefighter's seatbelt is known. This number does not include seatbelt use on watercraft, aircraft, or ATV's.

- Five firefighters were killed in crashes while wearing their seatbelts.

Additional information about firefighter deaths in 2005 as the result of vehicle crashes includes the following:

- Three firefighters who were employees of a private aircraft firefighting company were killed in the crash of their aircraft during a training exercise in California.

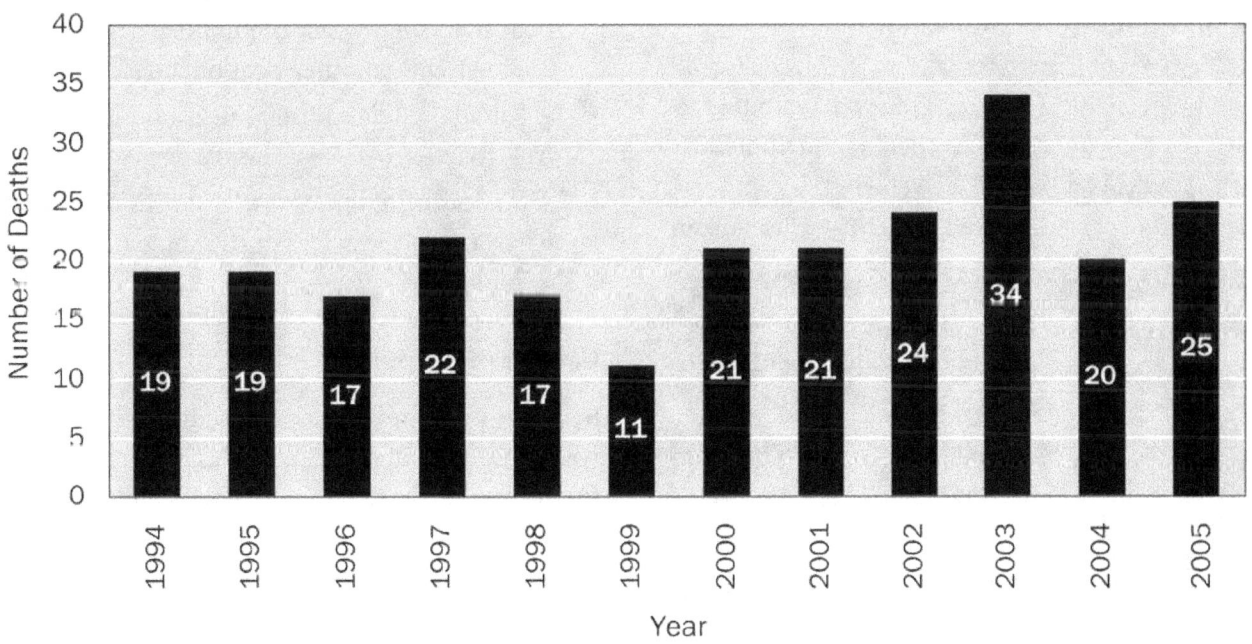

Figure 10. Firefighter Fatalities in Vehicle Collisions

- Three wildland firefighters were killed in the crash of a helicopter during a prescribed burn in Texas. The crew had been assigned to drop flammable spheres to initiate the burn when they experienced an unknown problem and crashed.

- Five firefighters were killed in crashes involving their vehicles:
  - Three crashes occurred while firefighters were responding to emergencies.
  - One crash occurred while the firefighter was returning from a class: A truck crossed the center line of the road and struck the firefighter's car.
  - One crash occurred as the firefighter drove from home to the fire station in response to an offshift callback.

The five personal vehicle crash deaths in 2005 bring to 64 the number of firefighter deaths in personal vehicle crashes since 1990. Many of these deaths involved excessive speed and lack of seatbelt use.

- Four fatal crashes involved the death of the driver of a fire department tanker (tender):
  - Two of the tanker crashes involved mechanical failure. One was attributed to a blowout and the other was attributed to a wheel falling off the vehicle.
  - One crash involved a tractor-trailer tanker.

- Four fatal crashes involved engine or pumper apparatus:
  - A California firefighter was killed when his engine company was involved in a crash in bad weather while responding to an incident.
  - A Georgia firefighter was killed in the crash of the engine that he was driving in response to a mutual-aid incident.
  - Two firefighters were killed in separate incidents when their apparatus were involved in crashes with tractor-trailer trucks.

- A New York firefighter was killed when the rescue truck he was driving back to station after an incident left the roadway and crashed.

- A Missouri firefighter was killed when the car he was driving was struck at high speed by the vehicle of an individual who was fleeing law enforcement.

- A Kansas firefighter was killed when the brush truck he was driving collided with a tanker responding to the same incident.

- A firefighter who was a passenger in a rescue boat was killed when the fire department boat collided with another boat on a dark waterway.

- A Nevada wildland firefighter was killed in an ATV rollover.

- A Missouri firefighter was run over by a push truck at a race track after the vehicle went out of control.

Caught or Trapped

In 2005, nine firefighters were killed when they were caught or trapped. This classification covers firefighters who are trapped in wildland and structural fires and unable to escape due to rapid fire progression and its byproducts of smoke, heat, and flame. This classification also includes firefighters who are killed in drownings.

- Two Wyoming firefighters were killed by rapid fire progress during an accidental fire in a townhouse. The firefighters were in the second story of the home searching for the fire when the fire rapidly emerged from a concealed space and trapped both firefighters.

- Two firefighters were trapped underwater and drowned in 2005. One firefighter was involved in a water rescue in Iraq and the other firefighter was engaged in water rescue training.

- Two firefighters died in basement fires. A New York firefighter was killed in the basement of a home when fire conditions developed rapidly, and a Pennsylvania firefighter was killed in a basement fire training scenario at a training facility.

- A Michigan firefighter was killed when fire progress blocked his escape from a residential structure fire.

- A Virginia firefighter was killed when rapid fire progress at a wildland fire overcame him.

- A Missouri firefighter was caught in machinery as he attempted to access the upper levels of a mill that was involved in a fire.

Falls

Five firefighters died in 2005 as a result of falls, the same number as were killed by falls in 2004. This was the fourth most common cause of fatal injury for firefighters in 2005.

- Two New York City firefighters were killed when they were forced to jump from a fifth-story window during a fire in a multiple-family residence.

- A Texas firefighter was killed when he fell from a ladder truck as it turned a corner during an emergency response.

Since 2002, eight firefighters have died as the result of falls from vehicles.

- A Connecticut firefighter was killed when he slipped and fell as he attempted to descend a ladder from the top of a training building.

- A Delaware firefighter died after a prolonged fight to overcome injuries received when he slipped and fell through a fire station pole hole.

Struck by Object

Being struck by an object was the fifth leading cause of fatal firefighter injuries in 2005. Four firefighters died after being struck by vehicles or other objects while on duty.

- A Texas firefighter was killed when a portion of a burning house collapsed and he was struck down and trapped in the debris.

- A North Carolina firefighter was killed when a large portion of a fire-damaged tree detached and fell on him, inflicting fatal injuries.

- A New Jersey fire police officer was killed when he was struck by a vehicle operated by a driver under the influence of alcohol.

- A Kentucky firefighter was killed when the tanker he had been driving rolled forward and crushed him.

Contact/Exposure

Three firefighters were killed in 2005 when they came into contact with or were exposed to harm:

- Two firefighters were electrocuted after coming into contact with an energized electrical wire.

- An Iowa firefighter was killed when he came into contact with manure gas while attempting to rescue a coworker in manure pit.

Assault

A Tennessee firefighter was killed when he was the victim of an assault. The firefighter was shot in a domestic violence situation between a secretary and her husband in the secretary's office, where the firefighter was waiting for a fire truck to be repaired.

Other

Six firefighters died in 2005 of causes that are not categorized above:

- A North Carolina firefighter died as a result of a heart valve problem.
- Two firefighters died of medication overdoses.
- Two firefighters died of pulmonary embolisms.
- A Florida firefighter died of heat stress.

NATURE OF FATAL INJURY

Table 10 and Figure 11 show the 115 deaths distributed by the medical nature of the fatal injury or illness.

Table 10. Nature of Fatal Injury (2005)

Nature	Number
Heart Attack	55
Internal Trauma	32
Asphyxiation	8
CVA	6
Burns	3
Crushed	2
Electrocution	2
Heat Exhaustion	1
Other	6
Total	**115**

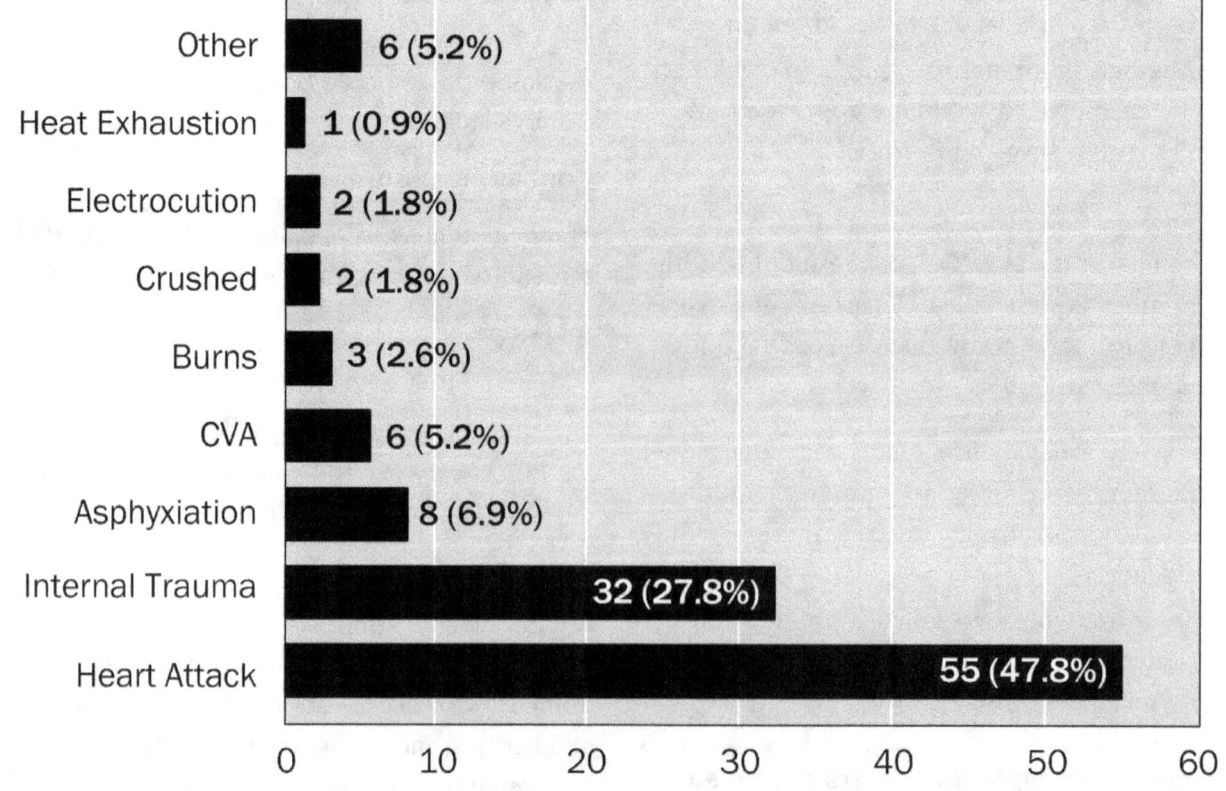

Figure 11. Fatalities by Nature of Fatal Injury (2005)

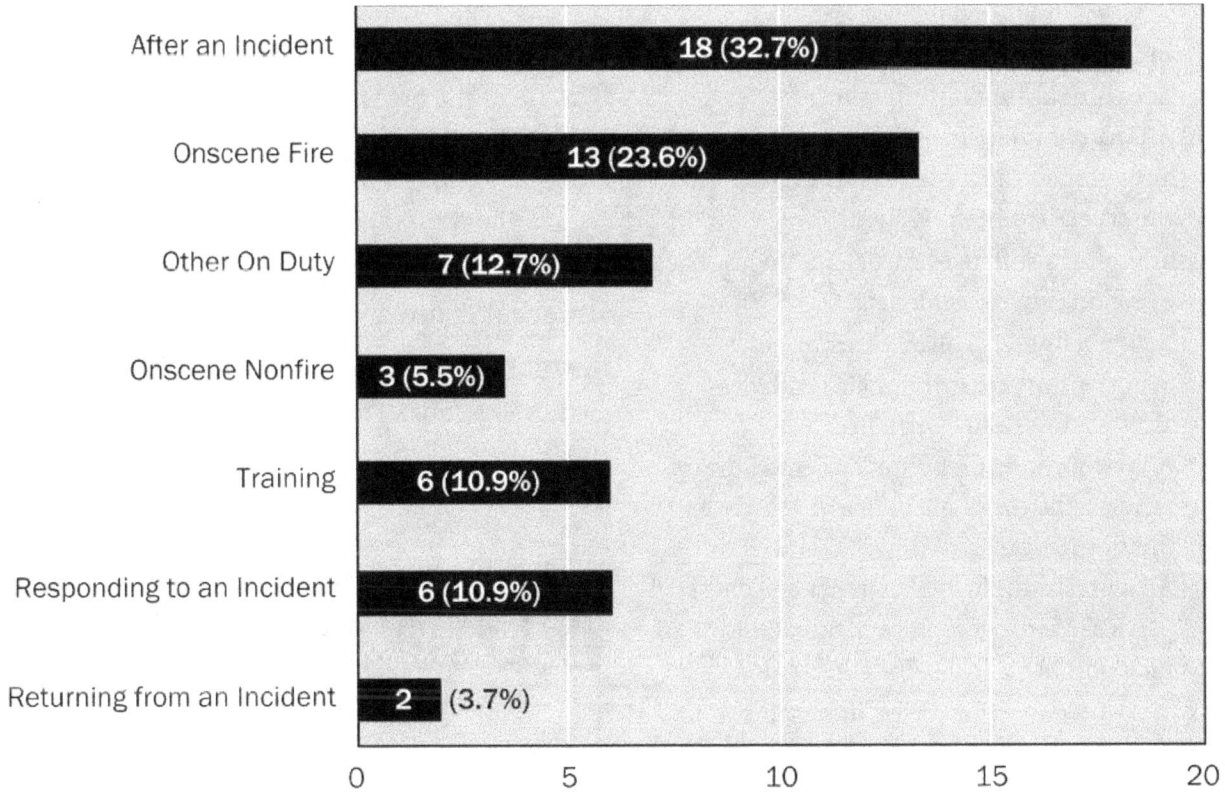

Figure 12. Heart Attacks by Type of Duty (2005)

Heart Attack

Heart attacks were the most frequent cause of death for 2005, with 55 firefighter deaths, down from 61 heart attack deaths in 2004. The timing of illness onset can range from the time at which the firefighter is walking from the station to his or her car after the conclusion of the incident, to the next day. Figure 12 provides a detailed breakdown of heart attacks by type of duty involved.

- Eighteen firefighters died of heart attacks that struck after the conclusion of an incident response or onduty period:
  - These firefighters suffered heart attacks within 24 hours of onduty stressful or strenuous activity.
  - Fourteen of the 16 firefighters included in this report under the Hometown Heroes criteria are in this group.

- Thirteen firefighters died of heart attacks that struck while they were working on the incident scene:
  - Three firefighters experienced heart attacks while fighting wildland fires.
  - An Arkansas firefighter suffered a heart attack while operating a fire pump at a car fire.
  - Eight firefighters were stricken with heart attacks at the scene of structural fires.
  - A Delaware firefighter collapsed of a heart attack while answering an automatic fire alarm in a building.

- Seven firefighters experienced heart attacks while on duty but not assigned to an incident or participating in training:
  - Four firefighters became ill as they worked or slept in the fire station while on duty.
  - A New York firefighter became ill as he marched in a fire department parade.

25

- A Mississippi firefighter suffered a heart attack as she participated in a fire department fundraising event.
- A Georgia firefighter suffered a heart attack and crashed the truck he was driving after the truck had been washed.

• Six firefighters suffered heart attacks while they were responding to incidents:

- Four firefighters collapsed from heart attacks prior to leaving their residences after having been dispatched.
- A West Virginia firefighter suffered a heart attack as he drove his personal vehicle to the fire station.
- A Nebraska firefighter experienced chest pains as he drove a piece of apparatus to an incident; he later died of a heart attack.

• Six firefighters were involved in training activities when they had heart attacks:

- Firefighters in Florida and Kentucky died while involved in SCBA training.
- Firefighters in New Mexico and New Jersey died while participating in physical fitness training.
- A New Mexico firefighter suffered a heart attack while participating in wildland tree-felling training.
- A Texas firefighter fell ill after training in the fire station.

• Three firefighters became ill and died of heart attacks that occurred while they were assigned to nonfire emergencies:

- A South Carolina firefighter died on the scene of an EMS incident.
- A New Jersey firefighter collapsed on the scene of a gas leak in a home, and later died.
- An Alabama firefighter arrived on the scene of a motor vehicle crash and collapsed shortly after getting out of his vehicle.

• Two firefighters suffered heart attacks as they returned from emergencies:

- A Kentucky firefighter suffered a heart attack as he drove his personal vehicle back to the fire station after responding to a lightning strike on a residence.
- A Georgia firefighter had a heart attack as he drove his engine company from the scene of a weather-related emergency.

Internal Trauma

In 2005, 32 firefighters died due to internal physical trauma. This grouping includes most firefighters killed in vehicle crashes and those who received physical injuries.

Table 11. Internal Trauma Firefighter Deaths

Year	Number of Firefighter Deaths
2005	32
2004	31
2003	41
2002	34
2001	28*
2000	36
1999	25
1998	27
1997	32
1996	32
1995	24
1994	21

* Does not include the firefighter deaths of September 11, 2001, in New York City.

• Two New York City firefighters were killed when they were forced to jump from a fire-involved structure.
• Three firefighters were killed in the crash of an airtanker during training.

- Three firefighters were killed when a helicopter crashed while supporting a prescribed burn in Texas.
- Five firefighters were killed in crashes involving their personal vehicles.
- Four firefighters were killed in tanker (tender) crashes.
- Four firefighters were killed in crashes that involved engine or pumper apparatus.
- Three firefighters died in falls. A Connecticut firefighter fell from a ladder, a Delaware firefighter slipped and fell through a pole hole, and a Texas firefighter died after falling from a ladder truck during a response.
- A New York firefighter died in the crash of a rescue truck as it returned to the fire station.
- A Kansas firefighter was killed when the brush truck he was driving crashed into a fire tanker that was responding to the same incident.
- A Missouri firefighter was killed when his car was struck by another car whose driver was fleeing law enforcement.
- A Missouri firefighter was killed when he was struck by a push vehicle while working a standby at a local race track.
- An Alabama firefighter was killed in a collision involving a fire rescue boat.
- A Nevada firefighter was killed in an ATV crash.
- A Tennessee firefighter died from gunshot wounds.
- A New Jersey fire police officer was struck and killed by a driver who was under the influence of alcohol.

Asphyxiation

Asphyxiation was the third leading medical reason for firefighter deaths in 2005, responsible for eight deaths:

- Four firefighters suffered fatal smoke inhalation in structure fires in Wyoming, Michigan, New York, and Houston.

- Two firefighters drowned, one in Pennsylvania during training and the other during a rescue attempt in Iraq.
- A Missouri firefighter was entangled in a man lift and died of positional asphyxiation.
- An Iowa firefighter died from exposure to manure gas during an attempted rescue.

Table 12. Firefighter Deaths due to Asphyxiation

Year	Number of Firefighter Deaths
2005	8
2004	5
2003	6
2002	15
2001	18
2000	13
1999	16
1998	15
1997	15
1996	5
1995	20
1994	29

Cerebrovascular Accident

Six firefighters died in 2005 as a result of strokes (CVA's):

- A New York firefighter suffered a CVA during structural firefighting training.
- A Tennessee firefighter suffered a CVA as he drove his engine back from a false alarm.
- An Oregon firefighter died of a CVA that struck at a fire management conference.
- A New Jersey firefighter suffered a stroke while driving a fire department pumper in a funeral procession.

- A Maryland firefighter was struck with a CVA when setting up for a fire department event.
- A New York firefighter collapsed due to a CVA shortly after arriving on the scene of a working structure fire.

Burns

Three firefighters died as a result of burns in 2005, the same number of firefighter deaths of this nature as in 2004.

- A Wyoming firefighter died of burns received in a structure fire when he and another firefighter were trapped by fire progress.
- A Pennsylvania firefighter died of burns received in a training exercise.
- A Virginia firefighter received fatal burns in a wildland fire.

Crushed

Two firefighters were killed in 2005 as a result of being crushed:

- A Kentucky firefighter was crushed by a tanker truck that he drove to the scene of a structure fire. The firefighter stopped the truck, walked in front of it, and was crushed as the truck rolled forward.
- A North Carolina firefighter was killed when a large portion of a fire-damaged tree fell on him.

Electrocution

Two firefighters were fatally electrocuted in 2005:

- A California firefighter was electrocuted when he came into contact with a live wire at a structural fire.
- Kansas firefighter was killed when he left his home and contacted an energized wire after reporting a wildland fire caused by lightning.

Heat Exhaustion

One firefighter was killed due to heat exhaustion in 2005. A Florida firefighter collapsed during a training run and subsequently died.

Other

Six firefighters died in situations in which the nature of their deaths does not fall into any of the categories described above:

- Firefighters in New York and Wisconsin died of pulmonary embolisms.
- Firefighters in Arizona and New York died of drug intoxication.
- A North Carolina firefighter died of a heart condition related to problems with heart valves.
- A Texas firefighter died of an aortic aneurysm.

FIREFIGHTER AGES

Figure 13 shows the percentage distribution of firefighter deaths by age and nature of the fatal injury. Table 13 provides counts of firefighter fatalities by age and the nature of the fatal injury.

As in most years, younger firefighters were more likely than older firefighters to die as a result of traumatic injuries, such as injuries from an apparatus accident or injuries sustained after becoming caught or trapped during firefighting operations. Stress plays an increasing role in firefighter deaths as age increases.

The youngest firefighter killed on duty in 2005 was Firefighter Trainee Justin Wisniewski of Connecticut. He died after a fall from a roof during training at age 18.

Table 13. Firefighter Ages and Nature of Fatal Injury

Age Range	Nontrauma Total	Trauma Total
under 21	0	3
21 to 25	0	5
26 to 30	2	3
31 to 35	2	5
36 to 40	7	7
41 to 45	9	5
46 to 50	11	5
51 to 60	22	10
61 & over	13	5

Note: Age of one firefighter unknown.

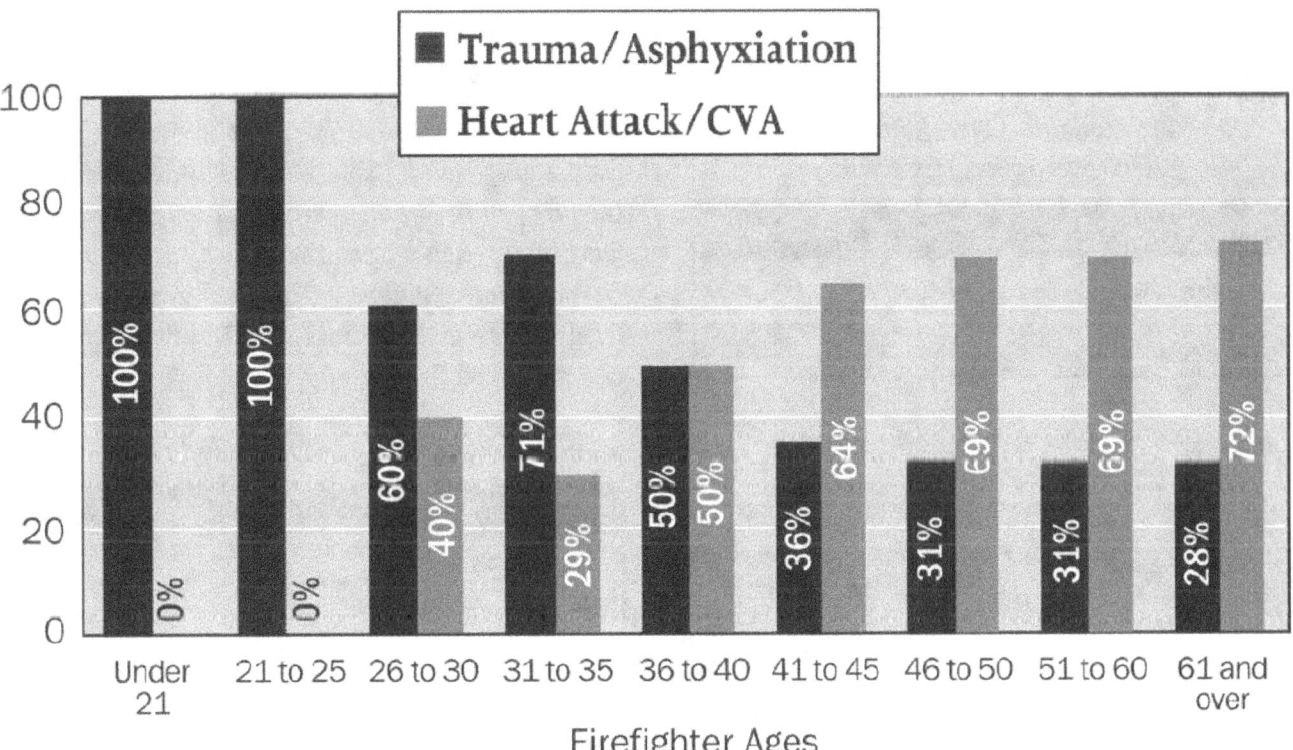

Figure 13. Fatalities by Age and Nature (2005)

The oldest firefighter killed on duty in 2005 was Fire Police Officer Joseph Walsh of New Jersey. He died at age 76 after being struck by a car driven by a drunk driver.

FIXED PROPERTY USE FOR STRUCTURAL FIREFIGHTING DEATHS

There were 19 firefighter fatalities in 2005 in which the firefighters became ill while on the scene of, or engaged in, structural firefighting. Table 14 shows the distribution of these deaths by fixed property use. As in most years, residential occupancies accounted for the highest number of these fireground fatalities, with 18 deaths.

Table 15 shows the number of firefighter deaths in residential occupancies for the last 8 years. Residential occupancies usually account for 70 to 80 percent of all structure fires and a similar percentage of the civilian fire deaths each year.* Historically, the frequency of firefighter deaths in relation to the number of fires is much higher for nonresidential structures.

Table 14. Structural Firerfighting Deaths by Fixed Property Use (2005)

Fixed Property Use	Number	Percent
Residential	18	94.7%
Commercial	1	5.3%

* Complete 2005 NFIRS fire incidence data were not available at the time of this report, but residential fires typically account for between 70 and 80 percent of all civilian fatalities each year, according to the NFPA.

Table 15. Firefighter Deaths in Residential Occupancies

Year	Number of Firefighter Deaths
2005	18
2004	15
2003	10
2002	21
2001	17
2000	21
1999	23
1998	17
1997	16
1996	19
1995	18
1994	25

TYPE OF ACTIVITY

In 2005, there were a total of 27 firefighter deaths on the fireground. Table 16 and Figure 14 show the types of fireground activities firefighters were engaged in at the time they sustained their fatal injuries or illnesses. This total includes all firefighting duties, such as wildland firefighting and structural firefighting.

Table 16. Type of Activity (2005)

Nature	Number
Fire Attack	11
Search and Rescue	6
Water Supply	2
Incident Command	2
Scene Safety	1
Other	5
Total	**27**

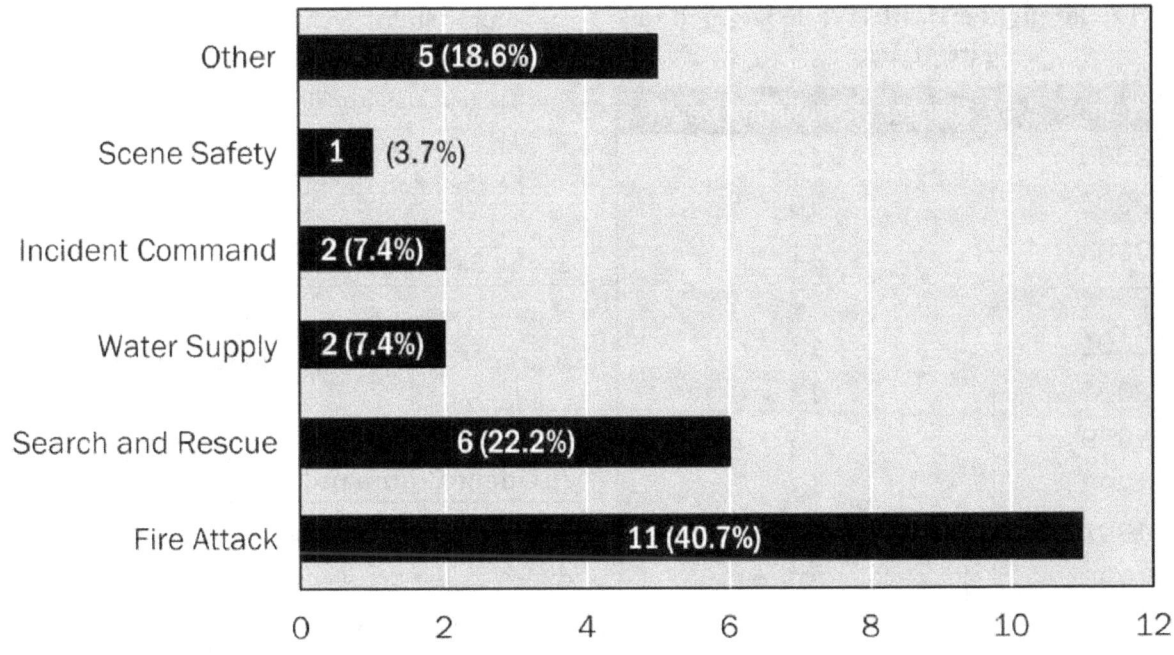

Figure 14. Fatalities by Type of Activity (2005)

Note: Onscene fire only, 27 of 115 fatalities.

Fire Attack

In 2005, 11 firefighters were killed as they engaged in direct fire attack, such as advancing or operating a hoseline at a fire scene. In years past, most fireground firefighter deaths occurred while the firefighter was engaged in fire attack (see Table 17).

- A Houston firefighter was killed when he was trapped by a structural collapse at a residential structure fire as he and his crew advanced an attack line to the fire.
- A Michigan firefighter helped to advance a hoseline into a residential structure, attempting to find the seat of a fire. He left the line to continue the search when fire conditions changed rapidly and trapped him in the structure.
- A California firefighter was electrocuted when he came into contact with an energized wire at a structure fire where his crew was assigned to exposure protection.

- A New Jersey firefighter was operating a handline on a defensive fire in a manufactured (mobile) home when he became ill; he later died of a heart attack.
- A California firefighter was helping to fight a fire in a historic home when he suffered a heart attack.
- A North Carolina firefighter was killed when a large fire-weakened tree branch fell on top of him and inflicted fatal injuries.
- An Oklahoma firefighter died of a heart attack that struck as he operated a handline on a wildland fire.
- A South Dakota firefighter was struck with a heart attack after he and other firefighters controlled a wildland fire caused by a fire in a hay bailer.
- A Mississippi firefighter collapsed suddenly after the control of a small wildland fire.
- A California firefighter collapsed from a heart attack after leaving a fire-involved structure to change his SCBA cylinder.
- A Virginia firefighter was overrun by fire progress and burned at a wildland fire.

31

Table 17. Firefighter Deaths While Engaged in Fire Attack

Year	Number of Firefighter Deaths
2005	11
2004	16
2003	11
2002	13
2001	13
2000	13
1999	16
1998	18
1997	21
1996	9
1995	18
1994	7

Search and Rescue

Six firefighters were killed in 2005 as they engaged in search-and-rescue activities:

- Two New York City firefighters were trapped by fire progress in a multiple dwelling as they searched for occupants. The firefighters were forced to jump out a fifth-story window due to fire conditions, and were killed in the fall.
- Two Wyoming firefighters were trapped by fire progress in a townhouse as they searched for children who were reported trapped in the structure.
- A New York firefighter collapsed shortly after completing the search of a house.
- A New York City firefighter was trapped by fire progress in the basement of a home as he performed a search.

Water Supply

Two firefighters died in 2005 while engaged in water supply duties:

- A Kentucky firefighter was crushed by the tanker (tender) he was driving in a water supply shuttle. The firefighter dismounted the truck, walked in front of it, and was struck by the truck as it rolled forward.
- An Arkansas firefighter suffered a heart attack as he operated the pump at a car fire.

Scene Safety

A Pennsylvania fire police officer suffered a heart attack while directing traffic at a structural fire in a residence.

Incident Command

Two firefighters died as they carried out command duties on the fire scene in 2005:

- A New York command officer collapsed from a CVA upon arriving at a working structure fire.
- An Arkansas fire chief felt ill and suffered a heart attack as he commanded a fire suppression operation.

Other Activity

Five firefighters became ill and later died of various injuries, not listed above, that were suffered on the fire scene:

- A New York fire coordinator suffered a heart attack on the fire scene.
- A Delaware firefighter drove an engine company to the scene of a fire alarm activation and collapsed upon arriving at the scene.
- A New York firefighter collapsed at the scene of a fire in his house after he and other firefighters had arrived on the scene.
- A Kansas firefighter was electrocuted as he went to investigate a wildland fire on his property caused by a lightning strike. He called the fire department, left the house, and was discovered by firefighters later in the evening.
- A Missouri firefighter was killed when he was entangled by a man lift as he ascended to the upper reaches of a mill to investigate a fire.

TIME OF INJURY

The distribution of all 2005 firefighter deaths according to the time of day when the fatal injury occurred, is illustrated in Figure 15. The time of fatal injury for seven firefighters either was not known or was not reported.

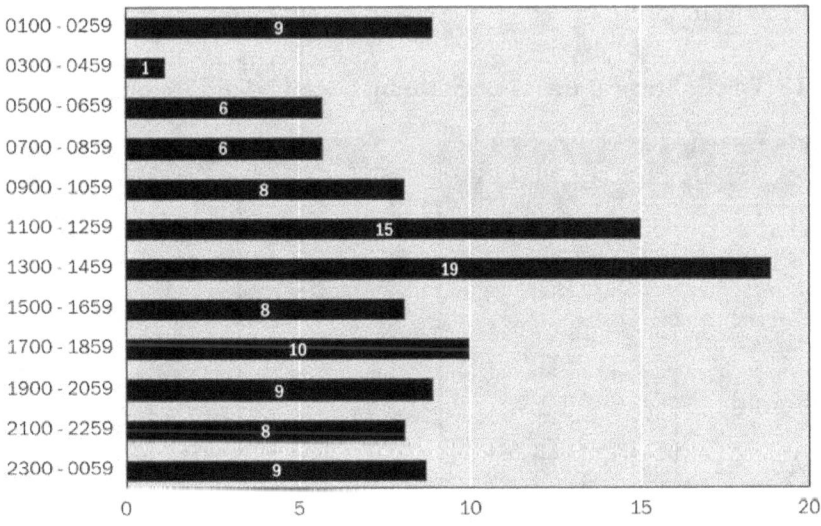

Figure 15. Fatalities by Time of Fatal Injury (2005)

MONTH OF THE YEAR

Figure 16 illustrates firefighter fatalities by month of the year.

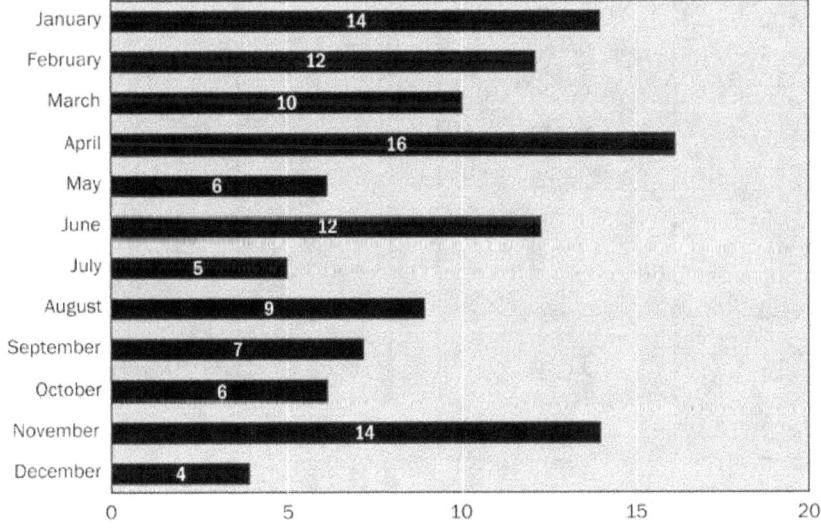

Figure 16. Deaths by Month of the Year (2005)

STATE AND REGION

The distribution of firefighter deaths by State is shown in Table 18. Firefighters based in 37 States died in 2005.

The highest number of firefighter deaths, based on the location of the fire service organization, occurred in New York, with 18 deaths in 2005.

Table 18. Firefighter Fatalities by State by Location of Fire Service* (2005)

Number	State	Percent of Total	Number	State	Percent of Total
4	Alabama	3.47%	1	Nebraska	0.86%
1	Arizona	0.86%	1	Nevada	0.86%
2	Arkansas	1.73%	1	New Hampshire	0.86%
9	California	7.82%	5	New Jersey	4.34%
2	Connecticut	1.73%	3	New Mexico	2.60%
2	Delaware	1.73%	18	New York	15.6%
3	Florida	2.60%	1	Oklahoma	0.86%
3	Georgia	2.60%	1	Oregon	0.86%
2	Indiana	1.73%	8	Pennsylvania	6.95%
2	Iowa	1.73%	1	South Carolina	0.86%
2	Kansas	1.73%	2	South Dakota	1.73%
6	Kentucky	5.21%	3	Tennessee	2.60%
1	Louisiana	0.86%	9	Texas	7.82%
2	Maryland	1.73%	1	Utah	0.86%
2	Michigan	1.73%	1	Virginia	0.86%
1	Minnesota	0.86%	2	West Virginia	1.73%
3	Mississippi	2.60%	1	Wisconsin	0.86%
3	Missouri	2.60%	2	Wyoming	1.73%
4	North Carolina	3.47%			

*This list attributes the deaths to the State in which the fire department or unit is based, as opposed to the State in which the death occurred. They are listed by those States for statistical purposes and for the National Fallen Firefighters Memorial at the National Emergency Training Center.

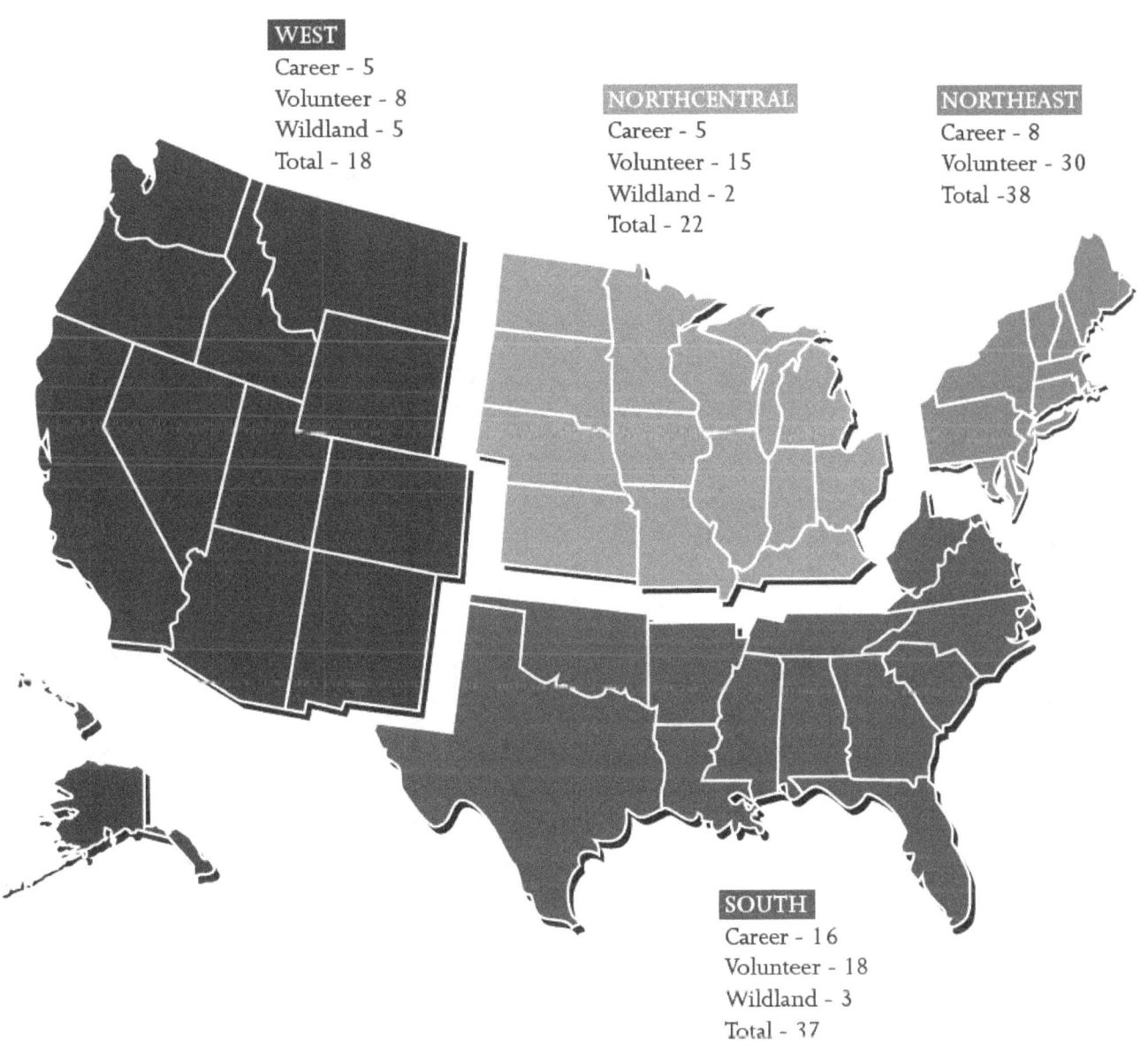

WEST
Career - 5
Volunteer - 8
Wildland - 5
Total - 18

NORTHCENTRAL
Career - 5
Volunteer - 15
Wildland - 2
Total - 22

NORTHEAST
Career - 8
Volunteer - 30
Total -38

SOUTH
Career - 16
Volunteer - 18
Wildland - 3
Total - 37

Figure 17. Firefighter Fatalities by Region (2005)

USA Onduty Firefighter Fatalities (2005)
Total: 115

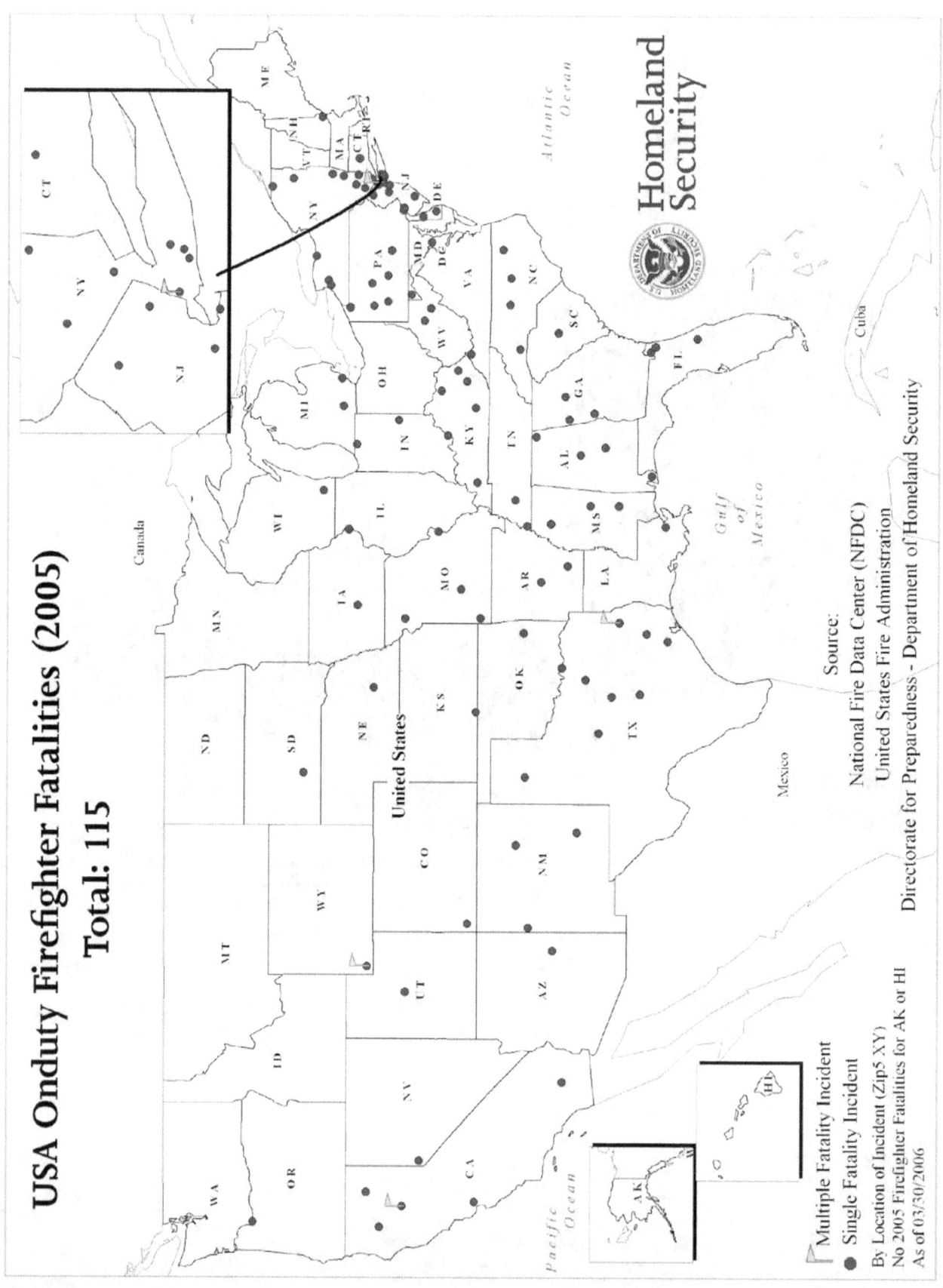

Multiple Fatality Incident
Single Fatality Incident
By Location of Incident (Zip5 XY)
No 2005 Firefighter Fatalities for AK or HI
As of 03/30/2006

Source:
National Fire Data Center (NFDC)
United States Fire Administration
Directorate for Preparedness - Department of Homeland Security

ANALYSIS OF URBAN/RURAL/SUBURBAN PATTERNS IN FIREFIGHTER FATALITIES

The United States Bureau of the Census defines "urban" as a place having a population of at least 2,500 or lying within a designated urban area. "Rural" is defined as any community that is not urban. "Suburban" is not a Census term, but may be taken to refer to any place, urban or rural, that lies within a metropolitan area defined by the Census Bureau, but not within 1 mile of the central cities of that metropolitan area.

Fire department areas of responsibility do not always conform to the boundaries used for the census. For example, fire departments organized by counties or special fire protection districts may have both urban and rural coverage areas. In such cases, it may not be possible to characterize the entire coverage area of the fire department as rural or urban, and firefighter deaths were listed as urban or rural based on the particular community or location in which the fatality occurred.

The following patterns were found for 2005 firefighter fatalities. These statistics are based on answers from the fire departments and, when no data from the departments were available, the data are based upon population and area served as reported by the fire departments.

Table 19. Firefighter Deaths by Coverage Area Type

	Urban/Suburban	Rural	Federal or State Parks/Wildland	Total
Firefighter Deaths	54	51	10	115

Information on chronic exposure to diesel exhaust for firefighters can be found at http://www.richter-foundation.org/home.html

IN CONCLUSION

In 2005, the fire service suffered another year of unacceptable loss; however, a number of hopeful signs emerged during the year.

- If the Hometown Hero firefighter deaths included in this study since 2003 are set aside for a moment, the number of firefighter deaths according to the standard report criteria (making multiyear analysis possible) dropped below 100 for the first time since 1998.

- The National Fallen Firefighters Foundation continued its work on preventing firefighter fatalities. The 16 initiatives that are the basis of their efforts were seen in fire service publications, included in fire service conference presentations, and included in local training. The Foundation's theme, "Everyone Goes Home," is pervasive.

- The IAFC sponsored a Firefighter Safety Stand Down. Nineteen fire service organizations partnered to make the event a success. This event encouraged fire departments to stop normal business for a day and devote that day or that shift to firefighter safety. According to the IAFC, an estimated 10,000 fire departments participated in 2005. The theme of the 2006 stand down will be vehicle safety.

The last firefighter fatality in 2005 was Firefighter/Paramedic Michael Hart of West Virginia. His car was struck by a tractor-trailer as he returned from teaching a class. The last firefighter death of 2004 also occurred in a vehicle crash.

- The IAFF, the IAFC, and the Volunteer Chief Officers Section of the IAFC joined forces to launch the National Fire Fighter Near-Miss Reporting System. This Web site allows firefighters to learn from the experiences of other firefighters. Lessons learned by firefighters who have almost been seriously injured or almost killed are posted on the site by the firefighters themselves. The site is available free of charge to all fire service members at www.firefighternearmiss.com/

- Federal fire service grant programs, recognizing the value of firefighter safety, have funded millions of dollars worth of firefighter safety equipment.

The task for 2006 and beyond is to determine whether these changes will have a lasting impact. Firefighter safety cannot be a fad or a temporary fix. These programs have pointed out that a cultural shift is required in the fire service—one that requires risk management at every emergency; and one that balances risks and benefits throughout the preparation for response and the response to emergencies.

PERSONNEL ACCOUNTABILITY

Saturday, February 19, 2005—0604hrs—Houston, Texas

Captain Grady Burke and the members of Houston Fire Department Engine 46 arrive 3 minutes after dispatch to find a working fire in a residence. The crew was ordered to execute a fast attack on the fire and stretched an attack line through the front door. Engine 46's crew was joined by ladder company firefighters as they moved into the interior of the structure to find the seat of the fire.

Without warning, a structural collapse occurred and fire advanced quickly into the areas occupied by the crew of Engine 46 and ladder company firefighters. An emergency evacuation was ordered by the Incident Commander (IC). With fire moving quickly and with great force, the firefighters inside the structure were forced to exit the house quickly through doors and windows.

Once they were outside, a Personnel Accountability Report (PAR) was ordered, and company officers and crews took head counts to make sure that everyone was out. As the PAR was being completed, an electronic accountability system used by the Houston Fire Department received a distress signal from Captain Burke's PASS device indicating that he was motionless.

A second alarm was ordered by the IC, and Rapid Intervention Crews (RIC's) were directed to the interior. Captain Burke was located under 2 to 3 feet of debris; he had expired as a result of smoke inhalation.

Firefighters operate in an extremely hazardous atmosphere when fighting structure fires. In addition to the dangers posed by the fire itself, firefighters often are forced to feel their way through a maze of rooms where they have never been before, in total darkness.

Changing fire conditions and unfamiliar surroundings lead firefighters to become lost or disoriented within structure fires. Many times, firefighters either cannot accurately describe their location to rescuers, or may describe their location incorrectly, or may be unable to communicate at all. Accountability systems assist firefighters and the IC in keeping track of the location and function of every firefighter on the emergency scene.

If something goes wrong on the fire scene, accountability systems give the firefighter his or her best chance for survival. This section will examine the requirements for accountability systems in incidents large and small.

Overview

Over the past 15 years, fire and emergency service departments have increasingly recognized the value of personnel accountability, for a variety of reasons. Maintaining complete awareness regarding the status of operating forces is essential for the safe and effective management of any fire or emergency scene. Accurate information on crew locations and assignments is needed for strategic and tactical decisionmaking, giving IC's and Division/Group/Unit Supervisors the assurance that critical tasks are performed with the required degree of coordination; this information also is important for tracking resource status, planning work-rest cycles, and carrying out demobilization activities. After a building collapse or other "mayday" situation, an updated list of personnel assignments will facilitate the identification, location, and rescue of lost, trapped, or endangered firefighters.

The need for continued attention to personnel accountability has, unfortunately, been bolstered by a series of firefighter line-of-duty deaths over the years. In some of these cases, personnel were "lost" in the chaos; in other examples, entire crews were mistakenly identified as being "safe," while their colleagues remained in danger.

Personnel accountability ultimately depends on more than any single system, technology, or procedure; it requires a combination of these, underpinned by a well-practiced incident management system and effective communications.

NFPA Standards 1500-2002 and 1561
There are two significant NFPA Standards that affect how fire departments create and implement their Personnel Accountability Systems—NFPA 1500 (Chapter 8—Emergency Operations) and NFPA 1561.

Incident Management

"An Incident Management System (IMS) is a management system used to organize emergency response and effectively and safely manage an incident."

No personnel accountability system can be expected to work in the absence of an established, properly functioning (standards-based) IMS doctrine. With the advent of the National Incident Management System (NIMS), there should be no circumstance in which firefighters are operating at a working incident without an IMS in place. The system alone is not enough, however; IC's, Safety Officers, and Division/Group Supervisors must take personal responsibility for knowing the whereabouts of operating personnel at all times. Furthermore, it is incumbent on system users to ensure that the system accurately reflects their location, crew, and work assignment at all times.

Communication

Personnel accountability ultimately rests on clear communication between individual firefighters and their supervisors at all levels in the IMS. Personnel must have the means and the will to notify the IC of any changes of their location or work assignment, so that the accountability system may be updated to reflect reality. Interoperability is also critical consideration; it would be difficult at best to maintain a personnel accountability system involving personnel on multiple radio channels without the ability to freely communicate through the IMS. Whatever the specific system used, the effectiveness of any personnel accountability system depends on communication between operating units and the accountability officer and/or IC.

Personnel Accountability Systems

Generally speaking, personnel accountability systems currently deployed by fire and emergency service departments in the United States for small-footprint events use an integrated set of components to identify functional units and personnel assigned to them. Helmet tags, replaceable helmet shields, "passports," hose tags, and other items are used to capture information, while a variety of accountability boards, pre-printed forms, collector rings, etc., are used to display the collected information in a format accessible to the IC and his or her subordinates. Regardless of the specifics of any given personnel accountability system, most systems use similar methods for determining how individuals and units are tracked throughout an incident.

Personnel Accountability Technology

Standards-based technology holds great promise for improving the state-of-the-art in personnel accountability; these advances generally fall into three categories: (1) technologies supporting the implementation of personnel accountability systems (e.g., hardware and software solutions for tracking and/or displaying personnel, units, and assignments); (2) individually deployed technologies improving the chance that downed firefighters or rescuers will be quickly located (e.g., integrated PASS devices, GPS transponders, firefighter rescue transceivers, etc.); and (3) technologies employing aspects of both (1) and (2).

A **PASS device** is an acronym for the **"Personal Alert Safety System,"** typically a one-way communications device used by firefighters entering a building to alert the outside Rapid Intervention Team (RIT) or Firefighter Assist and Search Team (FAST) that the wearer of the PASS device is in trouble and in need of rescue.

The PASS device has two methods of being activated (if turned on): manually by a firefighter or automatically when the firefighter is no longer moving. The device creates a high-pitched squealing or beeping sound, to be heard by fellow firefighters. These style devices are powered by battery and can be turned on using gloves and are safe to operate in flammable or explosive atmospheres.

Older versions of PASS devices were worn on the firefighter's belt or bunker gear, and they required the firefighter to turn them on separately from the SCBA. Because of this, there have been many fatalities in which firefighters were wearing their PASS devices, but the devices were not activated. Integrated PASS devices are now required to meet NFPA Standard 1981. However, there are still many SCBA in use across the United States that do not have integrated PASS devices, and thus are dependent on activation by the individual firefighter to work properly.

The future holds great promise for supporting personnel accountability systems through "next generation" devices that will track firefighters' whereabouts, biometrics, etc., throughout an incident. A standards-based approach, however, is a critical component of interoperability for these systems so that for mutual-aid and/or wide-area disaster responses, individuals, regardless of the specific vendor or technology employed, and the equipment they use will show up on and work

Currently, in major metropolitan areas such as Phoenix, Arizona, and Houston, Texas, the fire departments employ personnel accountability systems such as these:
Houston Fire Department—Houston uses a dual system of an integrated PASS and SCBA and a backup T-PASS device. The T-PASS (electronic accountability system) sends a radio signal to a base receiver that is maintained and monitored by the IC. The base receiver provides the IC with an audible and visual status of each firefighter on the scene.
The Phoenix Fire Department has used the passport accountability system for over a decade. All fire departments in the Phoenix metropolitan area use the same system. The system consists of small passport tags in each apparatus with the names of firefighters assigned to the apparatus attached by means of velcro to the passport. A standard system exists for the collection and use of the passports at small and large incidents. The names of firefighters assigned to a crew can be displayed on any mobile computer attached to the dispatch system.

with the system employed by any Safety Officer, anywhere. Some systems may provide a visual display to the IC at all times, helping to provide a common operating picture, maintain accountability, and reinforce the overall IMS by strengthening strategic and tactical decisions. Existing systems that

are available and in common use in the United Kingdom provide an automated "tally" (or list) of all members operating on an incident. Such systems also can support improved air management and work-rest cycles.

Thanks to members of the fire service who submitted Firefighter Field Experience Debriefing forms from their experiences working Katrina (Community Relations), many of which addressed the lack of functional command and control and the lack of personnel accountability.

Personnel accountability, an integral part of the Incident Command System (ICS), is most often thought of as being required only on discrete fire-ground locations; however, these technologies, systems, and policies also are needed on wildland fires, search and rescue incidents, and large-scale disasters such as Hurricane Katrina. These types of incidents most often cover large areas and increase the magnitude and complexity of personnel accountability. Fire service doctrine for accountability in such wide-area scenarios, especially when firefighters are not performing traditional functions, such as community relations, is in need of further consideration and development.

A universal map referencing system standard (a common map grid), critical at the human interface, be it voice communications protocols, paper maps, tracking charts, digital display, or "next generation" location-based technologies, is the U.S. National Grid (USNG-NAD83). USNG is a nationally defined map coordinate system based upon and interoperable with UTM (Universal Transverse Mercator) and the Military Grid Reference System (MGRS-WGS84), "a language of location," to supplement street

address and other local references for a variety of operational purposes (search on "grid" http://www.ci.lincoln.ne.us/CITY/FIRE/usar/members/misc/katrina/aar1.pdf), including a standards-based approach to personnel accountability systems and supporting Geospatial Information Technologies (GIT). In a wide-area scenario, for deployed individuals/teams from multiple agencies and jurisdictions operating under Unified Command, a concept both more important and more complicated when local, State, and Federal commanders are required to coordinate their efforts together (often in areas devoid of street signs), for safety's sake, knowing precisely where all responders are at all times, in case things go wrong and a Mayday is called, is critical. The USNG facilitates coordinated rapid intervention capabilities coming from disparate agencies/organizations in the area of operations.

The USFA recommends USNG as a standard referencing system at the user interface in order to ensure that everyone (fire service, law enforcement, military, and civilians) is on the same map page in the coming years (http://firechief.com/news/national-grid-usng8376/index.html).

For more information on the USNG go to http://www.fgdc.gov/usng/index.html

Community of Practice for GIT in Disaster and Emergency Management: http://mississippi.deltastate.edu/

FEMA Director R. David Paulison in written testimony submitted to the U.S. Senate Homeland Security and Governmental Affairs Committee for his confirmation hearing 05/24/2006:
"The NIMS Integration Center is considering the adoption of the 'National Grid' unified mapping system as a potential NIMS implementation standard. The mapping system would help saves lives, reduce the costs of the disaster, and enhance all disaster related actions."

Quick Fix—Safe Convoy Movements of Fire Service Personnel

From Katrina Firefighter Field Experience Debriefings regarding convoys under escort of law enforcement entities:

"We then ran back and proceeded to convoy down at 95 MILES PER HOUR!!! This was not only unsafe, it was unnecessary."

"Imagine 60 cars traveling on the Interstate between 95 and 100 mph—dodging debris and other cars. It certainly didn't make any sense to me (it wasn't like the hurricane had just occurred). It was not surprising that by the time we made it through the road blocks on the west side of the city, our convoy was scattered to the winds."

1. Discuss basic doctrine for convoy movements with the escort commander before departing, stressing safety considerations such as operating at a reasonable speed in consideration of all types of vehicles in the convoy (large trucks, high-profile multipac vans, etc.) and road conditions.

2. Have a basic operational plan and briefing for the movement that includes clear safety instructions (maximum safe speed, mandatory use of seatbelts!); properly gridded maps* (and GPS* if available) for each vehicle with the route and other necessary information clearly marked; other relevant directions and information regarding the movement provided to all vehicle operators before departing (grid designations and other addressing information for final destination, waypoints/turnpoints, traffic control points (TCP's) if known, special hazard areas, etc.); protocols to be followed (convoy sweep contingencies) should a vehicle encounter an emergency while en route, breakdown, or fall out of convoy for any other reason.

3. **At any time**, do not operate your vehicle, or allow a vehicle in which you are riding, to operate in an unsafe manner. If you consider a convoy movement in which you are a participant to be unsafe, indicate to other vehicles your intention to leave the convoy (turn signal, hazard signal lights) then do so in a safe manner.

4. **Report unsafe practices to your chain-of-command.**

Map grid and GPS both set to USNG-NAD83 or interoperable MGRS-WGS84.

SUMMARY OF 2005 INCIDENTS

January 2, 2005—1747hrs
Ornell Edgar Fuller, Jr., Captain
Age 40, Volunteer
Midway Volunteer Fire Department, Dexter, New Mexico

Captain Fuller and the members of his fire department responded to a structure fire in a single-family residence. Heat from the home's wood-burning stove had ignited the ceiling above the stove. The fire was controlled by firefighters, and all units were back in service by 1820hrs.

Captain Fuller went to work the next day. When he failed to return home that evening, his parents searched for him and found him dead at his place of work. Captain Fuller's death was caused by a heart attack.

Captain Fuller's father is the chief of the Midway Volunteer Fire Department.

January 3, 2005—2315hrs
Carl E. Sherman, Firefighter
Age 66, Volunteer
Southington Fire Department, Connecticut

Firefighter Sherman participated in a monthly fire department drill. The drill involved SCBA training activity at the fire station from 1930hrs through 2110hrs. The training consisted of repeatedly donning and doffing SCBA's, and of basic maintenance procedures. The training session ended and firefighters returned to their homes.

At 2315hrs, an ambulance was called to the home of Firefighter Sherman. Firefighter Sherman's wife had found him unresponsive in his bed. He was

transported to the hospital and admitted to the emergency room. He was diagnosed as suffering from cardiac arrest and was pronounced dead at 0025hrs on January 4, 2005.

January 6, 2005—0744hrs
Christopher R. DeWolf, Lieutenant
Age 40, Career
Newington Fire and Rescue, New Hampshire

Lieutenant DeWolf was at home, off duty. Due to weather conditions and incident activity, his shift was paged to respond to the fire station for coverage.

Lieutenant DeWolf was in uniform and driving his personal vehicle when he was involved in a single-vehicle crash. His vehicle lost traction on the snow-covered roadway, left the roadway, rolled over the guardrail, and collided with a sign pole and concrete base.

The crash caused major intrusion into the passenger area of Lieutenant DeWolf's Dodge Durango. The crash investigation indicated that Lieutenant DeWolf (who was wearing his seatbelt) was killed instantly as a result of severe head trauma. Unsafe speed for the road conditions was cited as the cause of the crash.

Lieutenant DeWolf had been a member of the Dover Fire Department for 17 years before joining Newington Fire and Rescue. He also was a well-known fire-service trainer.

January 7, 2005—1724hrs
Timmy Young, Fire Equipment Operator
Age 41, Career
Columbia Fire Department, South Carolina

Fire Equipment Operator Young and his engine crew responded to an EMS call at a residence. Firefighters entered the residence and found an individual who had been dead for some time. After exiting the dwelling, Fire Equipment Operator Young became ill and collapsed. Responders on the scene provided immediate care, and Fire Equipment Operator Young was transported to the hospital, suffering from a heart attack.

After undergoing emergency procedures at the hospital, Fire Equipment Operator Young was placed in the Cardiac Intensive Care Unit. He died on January 20, 2005, from complications related to the heart attack.

For additional information regarding this incident, please refer to NIOSH Fire Fighter Fatality Investigation and Prevention Program report F2005-06 (http://www.cdc.gov/niosh/fire/reports/face200506.html).

January 9, 2005—0600hrs
Robert Dewey Martin, Firefighter
Age 26, Volunteer
Bostic Volunteer Fire Department, North Carolina

On the evening of January 8, 2005, Firefighter Martin participated in a 2-hour EMS training session at the fire department. The training involved the movement of a 150-pound mannequin. Firefighter Martin did not complain of feeling ill during the training. When the session was completed, Firefighter Martin went home.

The next morning at approximately 0550hrs, Firefighter Martin's wife was awakened by the sound of Firefighter Martin struggling to breathe. A neighbor called 9-1-1 and provided CPR to Firefighter Martin. Paramedics arrived and provided ALS-level care at the scene and in the ambulance while en route to the hospital. Despite their efforts, Firefighter Martin was pronounced dead at the hospital.

At autopsy, a heart valve problem was discovered—a problem which may have been hereditary. Firefighter Martin did not suffer a heart attack.

For additional information regarding this incident, please refer to NIOSH Fire Fighter Fatality Investigation and Prevention Program report F2005-22 (http://www.cdc.gov/niosh/fire/reports/face200522.html).

January 11, 2005—1130hrs
Jerry Wayne Hopper, Forestry Technician
Age 61, Wildland Full-Time
Tennessee Department of Agriculture—Forestry Division

Forestry Technician Hopper was waiting for an apparatus to be repaired at the Tennessee Department of Transportation (TDOT) facility outside of Jackson.

Forestry Technician Hopper was waiting in an office area with three other people. The husband of a TDOT employee who worked in the office entered the room, drew a 9-mm pistol, and fatally shot his wife. The shooter then turned his gun on Forestry Technician Hopper, who was the person closest to the first victim, and shot him multiple times. Two others in the office were also shot, and a man outside the building was shot and killed.

The shooter left the facility and was captured later that day.

January 20, 2005—0730hrs
Scott Allen Thornton, Captain
Age 39, Career
Summit Township Fire Department, Michigan

Captain Thornton was in command of an engine company that was dispatched with other units to a report of smoke in a garage. Upon their arrival, Captain Thornton reported nothing showing. He made contact with the homeowner and found that the garage was charged with smoke.

Firefighters, including Captain Thornton, advanced an attack line into the structure and began to search for the source of the smoke. After difficulty in finding the fire, Captain Thornton and another firefighter began to search the upper floor of the residence without a handline. Captain Thornton's low-air alarm activated.

After firefighters had been on scene for approximately 20 minutes, fire conditions deteriorated rapidly and trapped Captain Thornton and the firefighter who had accompanied him. Captain Thornton ran out of air and buddy-breathed with the firefighter. A second attempt to buddy-breathe was unsuccessful, and Captain Thornton fell to the floor. The firefighter broke out a window, signaled distress, and went back to find Captain Thornton. The firefighter was unable to locate Captain Thornton, but was able to find his way out of the structure.

continued on next page

Approximately 20 minutes after his last radio transmission, Captain Thornton was found by firefighters and removed from the structure. Despite efforts on the scene and at the hospital, Captain Thornton was pronounced dead as a result of smoke inhalation.

Captain Thornton's air supply was depleted; his PASS device did operate, but the sound was muffled by the position of his body. The fire was caused by spontaneous combustion of oil-soaked rags that had been stored in the basement of the home.

For additional information regarding this incident, please refer to NIOSH Fire Fighter Fatality Investigation and Prevention Program report F2005-05 (http://www.cdc.gov/niosh/fire/reports/face200505.html).

January 20, 2005—0900hrs
Walter Matthew "Matt" Sarnoski, Firefighter
Age 19, Volunteer
Sandy Township Fire Department, Sabula Fire Station 39, DuBois, Pennsylvania

Firefighter Sarnoski and the members of his fire department were dispatched to a motor vehicle crash. Firefighter Sarnoski was operating his personal vehicle, a 1996 Plymouth Breeze.

For reasons unknown, Firefighter Sarnoski's vehicle left the right side of the roadway, traveled up a berm, returned to the roadway, crossed both lanes of traffic, and collided with a command vehicle responding to the same incident. After the collision, Firefighter Sarnoski's vehicle left the roadway and sustained extensive damage.

The drivers of both vehicles were wearing seatbelts. Firefighter Sarnoski was pronounced dead at the scene; the chief operating the command vehicle was transported to the hospital, treated for a concussion, and released.

January 23, 2005—0252hrs
Michael D. Falkouski, Assistant Fire Chief
Age 59, Career
Rensselaer Fire Department, New York

Rensselaer Fire Department units were dispatched to a report of a structure fire. Upon arrival, firefighters found a fully involved detached garage with a burn victim, and they requested a second alarm.

Assistant Chief Falkouski arrived on the scene in his personal vehicle and collapsed immediately upon exiting the vehicle. Firefighters on the scene started CPR and applied an AED.

Assistant Chief Falkouski was treated by paramedics who arrived approximately 15 minutes after his collapse. His treatment was continued in the ambulance en route to the hospital. Assistant Chief Falkouski was pronounced dead at the hospital. The cause of death was a CVA.

January 23, 2005—0815hrs
Curtis W. Meyran, Lieutenant
Age 46, Career
Fire Department New York, New York

John Gerard Bellew, Firefighter
Age 37, Career
Fire Department New York, New York

Lieutenant Meyran and Firefighter Bellew were assigned to Ladder 27 in the Bronx. At 0758hrs, Ladder 27 was dispatched to a structure fire in an occupied multiple-family dwelling.

Upon their arrival, Ladder 27 was assigned to provide ventilation, entry, and search services on the fourth floor of the building, above the fire. The fire progressed rapidly, and firefighters were trapped in an apartment that did not have a fire escape.

Six firefighters were forced to jump from the rear window into a depressed rear yard of the structure—in effect, a five-story fall. Two firefighters died and four firefighters received severe injuries. Blunt trauma injury was cited as the cause of death for both of the deceased firefighters.

A number of factors contributed to the deaths of the two firefighters. The building in which the fire occurred had been illegally converted into small apartments, making access and egress difficult. A frozen fire hydrant and the consequent loss of water supply to an attack line forced the movement of an attack line at the same time that the fire was extending rapidly. Snow was falling at the time of the fire, delaying unit arrivals and making movement difficult. The cause of the fire was electrical: an overheated extension cord.

At the time of the fire, FDNY firefighters were not equipped with escape ropes. Subsequent to the fire, all FDNY firefighters were equipped with escape and descent devices.

Firefighter Bellew was posthumously promoted to Lieutenant.

January 23, 2005—1347hrs
Richard Thomas Sclafani, Firefighter
Age 37, Career
Fire Department New York, New York

Firefighter Sclafani was assigned to Ladder 103 in Brooklyn. At 1337hrs, Ladder 103 and other fire department units were dispatched to a fire in a residence.

Firefighter Sclafani was a part of the Ladder's inside team, assigned to perform a search of the basement using the interior stairs. Fire conditions changed rapidly and Firefighter Sclafani became separated from other firefighters. He was unable to escape the basement.

Upon realizing that Firefighter Sclafani was missing, firefighters returned to the basement and removed him to the exterior. Firefighter Sclafani suffered third-degree burns to 63 percent of his body. His blood carbon-monoxide level was 24 percent. His cause of death was listed as smoke inhalation and burns.

The fire was caused when combustibles were placed in close proximity to an electric heater.

January 26, 2005—1200hrs
Donald Conner, Firefighter
Age 74, Volunteer
Brooken Volunteer Fire Department, Stigler, Oklahoma

Firefighter Conner was the only firefighter on the scene of a wildland fire. He was operating a 1-inch hoseline supplied by a brush truck. As he returned to the apparatus to reposition it, he became ill and collapsed.

Firefighters arriving on the scene witnessed the collapse and provided CPR. Firefighter Conner was transported to the hospital, where he later died. The cause of death was listed as a heart attack.

January 31, 2005—1430hrs
Walter Ray Minich, Driver
Age 63, Volunteer
Shermans Dale Community Fire Company, Pennsylvania

Driver Minich and the members of his fire department were dispatched to a medical emergency involving a child in respiratory arrest. Prior to the dispatch, Driver Minich had been working on a motor home.

Other firefighters noticed that Driver Minich had not responded to the incident. Minich was found at home, keys in hand, apparently having become ill while responding. He was transported to the hospital, where he was pronounced dead.

The cause of death was listed as a heart attack.

February 5, 2005—0100hrs
William "Bill" Matt Goodin, Captain
Age 56, Volunteer
Mount Victory Fire Department, Somerset, Kentucky

Captain Goodin and other members of his fire department had just completed their response to a medical assistance call. Most members departed for home, leaving Captain Goodin and the Fire Chief in the fire station.

Captain Goodin and the Chief left the fire station and Captain Goodin locked the door. He turned to the Chief and told him that he was not feeling well. Captain Goodin attributed his illness to heartburn, took two steps, and fell straight forward, striking his head on a vehicle bumper.

The Fire Chief called for assistance, and Captain Goodin was transported to the hospital. He was pronounced dead at the hospital; the cause of death was a heart attack.

February 6, 2005—1155hrs
Todd Raymond Smith, Firefighter
Age 31, Volunteer
New Paltz Fire Department, New York

Firefighter Smith and the members of his fire department were dispatched to the scene of a report of smoke in the basement of a residential structure. Firefighter Smith responded to the fire station and awaited the arrival of other firefighters to staff the apparatus.

A Chief Officer responded directly to the scene and found that there was no emergency. The Chief ordered all equipment to remain in-station and then to go back in service. Firefighter Smith left the station, and had begun to walk toward his vehicle when he experienced an unwitnessed collapse.

continued on next page

Approximately 36 minutes after the incident was concluded, a civilian driving by the fire station noticed Firefighter Smith lying in the parking lot. Despite treatment provided at the scene, in the ambulance, and at the hospital, Firefighter Smith died.

The medical examiner concluded that Firefighter Smith died due to drug intoxication, but did not rule out the possibility of a heart attack. Firefighter Smith had been prescribed pain medications for previous back injuries.

For additional information regarding this incident, please refer to NIOSH Fire Fighter Fatality Investigation and Prevention Program report F2005-24 (http://www.cdc.gov/niosh/fire/reports/face200524.html).

February 10, 2005—1420hrs
William Wade Pierce, Firefighter
Age 53, Volunteer
Ogdensburg Fire Department, New Jersey

Firefighter Pierce and members of his department responded to a mutual-aid gas explosion and fire in a manufactured home. The fire was caused when a propane delivery vehicle failed to properly disconnect the hose connection between the vehicle and the storage tank that served the residence. Upon the arrival of firefighters on scene, Firefighter Pierce started operating a handline while performing an exterior attack on the fire.

Firefighter Pierce began to feel unwell and went to his apparatus to request help from the apparatus operator. Firefighter Pierce was treated on the scene by EMT's and paramedics, and then transported to the hospital. Despite these efforts, he was later pronounced dead at the hospital. The cause of death was listed as a heart attack.

February 12, 2005—1819hrs
Angelo Petta, Chief Engineer
Age 46, Volunteer
City of Garfield Fire Department, New Jersey

Chief Engineer Petta and the members of his department were dispatched to a report of a gas leak in a residence. Since his residence was a short distance from the incident, Chief Engineer Petta responded to the scene in his personal vehicle.

Upon his arrival at the scene, Chief Engineer Petta entered the home and began to evacuate the residents. He was joined inside by the first-arriving command officer. Once the residence was evacuated, both firefighters left the residence to meet the other arriving firefighters. After stepping down from the porch in the front of the residence, Chief Engineer Petta suddenly collapsed.

Members of the fire and police departments provided immediate medical care, including the use of a defibrillator. Chief Engineer Petta was transported to the hospital, but was pronounced dead shortly after arrival. The cause of death was a heart attack.

February 13, 2005—0325hrs
Mark Francis "Mac" McCormack, Captain
Age 36, Career
Santa Clara County Fire Department, California

Captain McCormack was in command of Engine 10. At 0220hrs, Engine 10 and other fire department units were dispatched to a report of a structure fire in a residence. The first-arriving fire department unit reported a working fire with smoke and flames visible from the roof, and with most of the second floor of the residence involved in fire. The first-arriving officer declared a defensive strategy for the fire attack.

When Engine 10 arrived on the scene, they were assigned to provide exposure protection to the rear of the involved structure. As the fire fight continued, a second alarm and a third were requested.

At approximately 0325hrs, Captain McCormack was walking around the structure, and came into contact with a live 12kV electrical wire. The live wire was moved away from Captain McCormack, and treatment was initiated. He was transported to the hospital, and was pronounced dead at 0414hrs.

After an investigation of the incident, the California Division of Occupational Safety and Health cited the Santa Clara County Fire Department for not having a procedure to deal with live wires on the incident scene, for failing to erect barriers around the hazard, and for failing to keep firefighters away from an energized electrical line.

For additional information regarding this incident, please refer to the Santa Clara County Fire Department report on the incident. The report is available on the fire department Web site at http://www.sccfd.org follow the "McCormack Investigation Summary" link on the site.

February 13, 2005—0515hrs
Ray Rangel, Staff Sergeant
Age 29, Career
U.S. Air Force, Dyess Air Force Base, Texas

Staff Sergeant Rangel was assigned to the Logistical Support Area Anaconda, 50 miles north of Baghdad, near Balad, Iraq.

Staff Sergeant Rangel and other firefighters were dispatched to an area west of Balad, where a Humvee and crew had entered a 50-foot-wide canal and overturned. Two soldiers were trapped inside.

Firefighters arrived by helicopter, and two firefighters immediately entered the cold water of the canal to attempt a rescue. Both firefighters were incapacitated by the cold water and began to drift downstream.

Staff Sergeant Rangel rushed down the embankment, wearing his ballistic vest. Soldiers watched as Staff Sergeant Rangel extended his hand to one of the firefighters and then entered the water. Staff Sergeant Rangel was weighted down by the armor plates in his ballistic vest and drowned. His body was recovered hours later, downstream.

February 15, 2005—1336hrs
Michael Lee Crawford, Lieutenant
Age 51, Career
Carroll County Fire Rescue, Georgia

Lieutenant Crawford was on duty at the fire station, and performed standard maintenance and administrative duties. At approximately 1315hrs, Lieutenant Crawford drove the squad truck out of the fire station, intending to air-dry it after it had been washed. A short distance from the fire station, Lieutenant Crawford became incapacitated; the squad truck went off the road and down a 40-foot embankment.

Witnesses to the incident called 9-1-1, and firefighters were dispatched to the scene. Firefighters found Lieutenant Crawford in the driver's seat, pulseless and not breathing. Lieutenant Crawford was removed from the squad and placed on a backboard. An AED was applied by firefighters, and shocks were delivered.

Paramedics arrived and continued care. Lieutenant Crawford was transported by ambulance to a local hospital. He was later flown to a regional hospital for advanced cardiac care. Lieutenant Crawford died of complications related to his heart attack on February 19, 2005.

For additional information regarding this incident, please refer to NIOSH Fire Fighter Fatality Investigation and Prevention Program report F2005-10 (http://www.cdc.gov/niosh/fire/reports/face200510.html).

February 17, 2005—2339hrs
Michael Angelo Mercurio, Jr., Firefighter/EMT
Age 52, Paid-on-Call
Urbandale Fire Department, Iowa

Firefighter/EMT Mercurio responded to two emergency incidents on February 17. The first incident was an EMS call involving a patient transport to the hospital. The EMS incident was concluded at 2304hrs. At 2339hrs, Firefighter/EMT Mercurio and his engine company responded to a vehicle fire.

The fire had been knocked down by a civilian with a fire extinguisher. Firefighter/EMT Mercurio, wearing full PPE, including SCBA, was responsible for opening the hood of the car to complete extinguishment. Due to damage from the fire, the process of opening the hood was complicated.

At the conclusion of the incident, firefighters returned to the station and placed their equipment back in service, and Firefighter/EMT Mercurio left the station to return home at approximately 0130hrs. He did not complain of any health problems at the time of the incident.

When Firefighter/EMT Mercurio failed to show up for work the next afternoon, a fire department ambulance was sent to his residence for a check welfare call. Firefighters found Firefighter/EMT Mercurio face-down in his bed, obviously deceased. The cause of death was listed as a heart attack.

February 19, 2005—0601hrs
Grady Don Burke, Captain
Age 39, Career
Houston Fire Department, Texas

Captain Burke's engine company and other fire department units were dispatched to a report of a structure fire in a residential occupancy. Captain Burke and his engine were the first fire department unit on scene, arriving 3 minutes after dispatch. They found a working fire in a single-story residence.

continued on next page

Captain Burke and the members of his company stretched an attack line and made entry into the front of the structure. Interior visibility was good in the front area, but decreased as firefighters advanced toward the kitchen area at the rear. Movement inside the structure was complicated by debris left in the building by vagrants.

Roof ventilation was not possible, due to fire conditions, and ladder company firefighters entered the interior of the structure. A second attack line was extended into the building.

At approximately 0610hrs, the roof of an addition at the rear of the structure collapsed into the rear portion of the building. The collapse caused rapid fire progress in the interior and forced firefighters to exit the building quickly. A standard fireground evacuation signal was sounded. A PAR was completed, and a second alarm was ordered. A RIT was directed into the structure to assist firefighters.

An electronic accountability system indicated that Captain Burke's PASS device was in alarm. The RIT began to search for him by following hoselines into the fire area.

Captain Burke was located at approximately 0629hrs, under 2 to 3 feet of debris. The cause of death was listed as smoke inhalation and thermal injuries. Captain Burke's blood carbon-monoxide level was 26 percent. The fire was caused by arson.

A report on this incident will be prepared by the Texas State Fire Marshal. The report will be available at http://www.tdi.state.tx.us/fire/fmloddinvesti.html

For additional information regarding this incident, please refer to NIOSH Fire Fighter Fatality Investigation and Prevention Program report F2005-09 (http://www.cdc.gov/niosh/fire/reports/face200509.html).

February 21, 2005—1924hrs
Henry DeAngelo Hobbs, Jr., Senior Forest Ranger
Age 38, Wildland Full-Time
Florida Division of Forestry—Jacksonville Division, Florida

Senior Forest Ranger Hobbs reported for work at approximately 0800hrs. He and another firefighter went to the sites of two previous wildland fires, and at each site performed overhaul and mop-up operations involving the use of handtools and hoselines. The firefighters returned to the station at approximately 1200hrs.

The afternoon hours were spent completing paperwork and doing physical fitness activities. Senior Forest Ranger Hobbs went off duty and drove to his residence at approximately 1700hrs. Prior to his departure, he complained of indigestion.

After his arrival at home, Senior Forest Ranger Hobbs took over-the-counter medications for heartburn. He went outdoors and began to talk with his neighbor. At approximately 1817hrs, Senior Forest Ranger Hobbs collapsed.

A paramedic ambulance arrived at 1831hrs and found Senior Forest Ranger Hobbs unresponsive. He was transported to the hospital, with medical care continuing during the transport. Senior Forest Ranger Hobbs was pronounced dead shortly after arriving at the hospital. The cause of death was a heart attack.

For additional information regarding this incident, please refer to NIOSH Fire Fighter Fatality Investigation and Prevention Program report F2005-21 (http://www.cdc.gov/niosh/fire/reports/face200521.html).

February 22, 2005—2345hrs
Lonnie Wayne Nicklas, Fire Chief
Age 39, Volunteer
Shepherd Volunteer Fire Department, Texas

Chief Nicklas participated in training on the evening of February 22, 2005, and complained of not feeling well before going home for the night. The next day he worked his normal job and returned to the fire station after work. He checked equipment at the station and then went out for dinner with his wife. After completing other tasks, Chief Nicklas went to bed at approximately 2230hrs.

At about 2330hrs, Chief Nicklas' wife awoke to find the Chief complaining of stomach pain and shoulder pain. She began to call 9-1-1, but the Chief insisted that she take him to the hospital. As they prepared to depart, the Chief collapsed. EMS was called and arrived at the home. An AED was applied and shocks were administered.

Paramedics arrived and continued care throughout the transport to a local hospital. After unsuccessful treatment at the hospital, Chief Nicklas was pronounced dead. The cause of death was listed as a heart attack.

For additional information regarding this incident, please refer to NIOSH Fire Fighter Fatality Investigation and Prevention Program report F2005-25 (http://www.cdc.gov/niosh/fire/reports/face200525.html).

February 27, 2005—0027hrs
Michael Alfred Aunkst, Firefighter
Age 45, Volunteer
Benedict Rural Fire Protection District, Nebraska

Firefighter Aunkst and members of his fire department were dispatched to provide mutual aid at a barn fire. Firefighter Aunkst responded to the fire station and drove a fire department tanker (tender) to the scene.

During the response, Firefighter Aunkst experienced abdominal discomfort and feelings of gas pressure. When he arrived on the scene, he drank a large quantity of water, and the pain changed to chest pain.

Firefighter Aunkst was treated at the scene and then transported by ambulance to a location where the ambulance was met by paramedics. The transport continued to the hospital. As Firefighter Aunkst was being treated at the hospital, he went into cardiac arrest. Despite extensive efforts by emergency room staff, Firefighter Aunkst was pronounced dead at 0227hrs. The cause of death was a heart attack.

March 3, 2005—1130hrs
Thomas Logan Mower, Fire Police Officer
Age 62, Volunteer
Goodwill Fire Company, Glenolden, Pennsylvania

Fire Police Officer Mower and the members of his fire department were dispatched to an automatic fire alarm in a home. As he proceeded to his vehicle to respond, Fire Police Officer Mower was struck with a heart attack, and collapsed by the side of his car.

EMS was summoned, and Fire Police Officer Mower was transported to the hospital. He was pronounced dead a short time later. The cause of death was listed as sudden cardiac death.

March 9, 2005—1610hrs
James E. Mero, Jr., Fire Investigator
Age 51, Volunteer
Essex County Office of Emergency Services, Elizabethtown, New York

Fire Investigator Mero was on the scene of a fully involved residential structure fire in his role as an Essex County deputy fire coordinator.

Fire Investigator Mero had taken some photographs of the scene as a part of his investigative duties, and had then returned to his vehicle to retrieve additional equipment. He suffered a heart attack while in his vehicle.

Firefighters and EMS workers on the scene provided care, and Fire Investigator Mero was transported to the hospital. Despite treatment provided on the scene, in the ambulance, and at the hospital, Fire Investigator Mero died as a result of a heart attack.

March 10, 2005—0843hrs
Gerald Joseph "Jerry" Buehne, Fire Chief
Age 64, Career
Affton Fire Protection District, Missouri

Chief Buehne was traveling to a meeting of fire officials in his fire department sedan, a 2000 Ford Crown Victoria.

As he drove, Chief Buehne's vehicle was struck at a high rate of speed by the vehicle of an individual fleeing law enforcement. The driver's side of Chief Buehne's vehicle received extensive damage, and the vehicle was overturned as a result of the impact.

Firefighters arrived and engaged in a difficult extrication. Chief Buehne was removed from his vehicle and transported to the hospital, where he was pronounced dead. The cause of death was multiple trauma. Chief Buehne was wearing his seatbelt at the time of the crash.

March 10, 2005—1354hrs
Charles Lynn Edgar, Fire Management Officer
Age 54, Wildland Full-Time
USDA Forest Service—Sabine National Forest, Texas

John Greeno, Heliport Base Manager
Age 51, Wildland Full-Time
USDA Forest Service—Stanislaus National Forest, California

José Victor Gonzales, Pilot
Age 45, Wildland Contract
Brainerd Helicopter Service Under Contract to the USDA Forest Service, Minnesota

continued on next page

Edgar, Greeno, and Gonzales were assigned aerial operations at a prescribed burn in the Sabine National Forest in Texas. The crew staffed a helicopter that was assigned to drop plastic ignition spheres from a low altitude.

During the first flight of the day, the sphere ejection machine jammed and the helicopter returned to base. The machine was repaired, and the helicopter and crew left to resume their mission.

Six minutes after their departure, a helicopter crew member reported that sphere firing was beginning. Two minutes later, a distress call was received from the helicopter. The aircraft crashed into a heavily wooded area and all three occupants were killed.

For additional information about this crash, consult the National Transportation Safety Board Web site at http://www.ntsb.gov/ntsb/query.asp—NTSB identification DFW05FA086.

March 16, 2005—1215hrs
Andre Miguel "Mike" Ellis, Sergeant
Age 39, Volunteer
Dixie Suburban Fire Department, Louisville, Kentucky

Sergeant Ellis and another firefighter had constructed a search-and-rescue training activity in an acquired structure. In order to establish that the activity was ready, and to establish a completion timeline, Sergeant Ellis and the other firefighter completed the activity.

The training activity consisted of donning full structural firefighting protective clothing, including SCBA; running up four flights of stairs; performing a 50-yard hose pull; and negotiating a search-and-rescue maze. When the firefighters had completed the activity, they sat down to rest for approximately 5 minutes.

When the firefighters rose from their rest, Sergeant Ellis suddenly collapsed. The firefighter working with Sergeant Ellis summoned EMS personnel from the base next door to the training area. Sergeant Ellis was treated at the scene and transported to the hospital. He was pronounced dead at the hospital after extensive medical efforts. The cause of death was listed as a heart attack.

For additional information regarding this incident, please refer to NIOSH Fire Fighter Fatality Investigation and Prevention Program report F2005-32 (http://www.cdc.gov/niosh/fire/reports/face200532.html).

March 21, 2005—2021hrs
Allen Wayne Wright, Fire Chief
Age 51, Volunteer
Hollywood Volunteer Fire Department, Alabama

Chief Wright and the members of his fire department were dispatched to a report of a vehicle crash. Chief Wright advised dispatch that he was responding from his residence.

A short time later, the dispatch center received a call for a possible heart attack at Chief Wright's address. Chief Wright was found in his vehicle, unresponsive. He was transported to a local hospital, but did not survive. His death was attributed to a heart attack.

March 26, 2005—0918hrs
Robert G. Brooks, Sr., Firefighter Trainee
Age 42, Volunteer
Montgomery Fire Department, New York

Firefighter Trainee Brooks was engaged in the final evolution of a Firefighter I course at the Orange County Fire Training Center. He was wearing full structural protective clothing, including SCBA. He was assigned to do a 360 of the training building, and prepare to go to the second floor and perform a search.

Firefighter Trainee Brooks completed the 360 and was about to attach his regulator to his facepiece when he suddenly collapsed. Instructors removed his gear and found that he was unresponsive. While awaiting the arrival of an ambulance, firefighters provided medical attention to Firefighter Trainee Brooks.

Firefighter Trainee Brooks was transported to the hospital and received further treatment. He died on March 28, 2005, as a result of a CVA.

March 29, 2005—1800hrs
Brandon Scott Phillips, Firefighter/Paramedic
Age 26, Career
Keller Fire Rescue, Texas

Firefighter/Paramedic Phillips went home after coming off duty from an overtime shift on March 29, 2005. During the shift, Firefighter/Paramedic Phillips ran three EMS calls that involved treatment and transportation of patients to the

continued on next page

hospital by ambulance. Firefighter/Paramedic Phillips also engaged in physically demanding duties while at the fire station.

Upon arriving home, Firefighter/Paramedic Phillips began his normal 5-mile physical fitness run. As Firefighter/Paramedic Phillips jogged, he felt a pop in his chest and began to have severe back pain. He went to a hospital where he was diagnosed as having an aortic dissection. He was transferred to a regional hospital for advanced care.

Firefighter/Paramedic Phillips underwent a 15.5-hour surgical operation to repair his heart. He did not recover from his surgery, and died on March 30, 2005.

April 2, 2005—1330hrs
Justin M. Wisniewski, Firefighter
Age 18, Volunteer
Southington Fire Department, Connecticut

Firefighter Wisniewski was participating in training at the Wolcott State Fire School. He was engaged in an exercise in which he was required to climb a ladder in full structural protective clothing and SCBA, dismount the ladder onto a roof, and climb down a ladder to the ground on the other side of the building.

Upon his arrival on the roof, Firefighter Wisniewski asked permission from an instructor to disconnect his regulator, since he had run out of air. The instructor gave him permission to disconnect and noted that Firefighter Wisniewski's facepiece was fogged.

As Firefighter Wisniewski mounted the ladder to descend, his foot slipped and he fell head first to the ground, a distance of approximately 20 feet. The instructor attempted to grab onto Firefighter Wisniewski's sleeve but his grip slipped. Firefighters provided immediate medical assistance and Firefighter Wisniewski was transported to the hospital. He died the next day as a result of his injuries.

April 2, 2005—2000hrs
William H. Poage, Jr., Firefighter
Age 51, Volunteer
Pintlala Volunteer Fire Department, Alabama

Firefighter Poage had just arrived at the scene of a rollover vehicle crash. He stepped out of his vehicle and collapsed of an apparent heart attack. Treatment was provided at the scene and he was transported to the hospital. Despite all efforts, he did not survive.

April 2, 2005—2045hrs
Phillip S. Young, Firefighter
Age 59, Volunteer
Robbins Hose Company No. 1, Dover Fire Department, Delaware

Firefighter Young had responded to multiple incidents on April 2, 2005. He remained in the fire station due to the approach of a storm. At approximately 2045hrs, he was preparing to descend from the second to the first floor by fire pole when he slipped.

Firefighter Young fell 12 feet to the floor below and sustained serious leg injuries. The wound failed to heal in a timely manner, and Firefighter Young was discovered at home with CVA-like symptoms on August 20. He was transported to the hospital, where he was diagnosed as suffering from a CVA. He died on August 22, 2005.

Firefighter Young was an administrator at the Delaware State Fire School for over 20 years, retiring in 2002.

April 11, 2005—1150hrs
James Michael Ratcliffe, Safety Officer
Age 62, Volunteer
Metuchen Fire Department, New Jersey

Safety Officer Ratcliffe served as a pallbearer and delivered the eulogy at the funeral for a former Fire Chief. As Safety Officer Ratcliffe drove a pumper in the funeral procession, he suffered a CVA. Safety Officer Ratcliffe died from complications associated with the CVA on May 3, 2005.

April 11, 2005—0851
Richard Allen Fast, Firefighter
Age 49, Volunteer
Midway Volunteer Fire Department, West Virginia

Firefighter Fast was driving a rescue truck to a motor vehicle crash. As he drove, Firefighter Fast suffered a heart attack and slumped over the steering wheel of the vehicle. The apparatus came into contact with guardrails on a bridge and stopped. Firefighters following in another vehicle stopped and provided assistance to Firefighter Fast.

continued on next page

An AED was applied, and paramedics arrived on the scene to continue care. Firefighter Fast was pronounced dead at the scene.

April 16, 2005—0200hrs
Dale A. Monica, Firefighter
Age 54, Volunteer
Burke Volunteer Fire Department, New York

A fire occurred in Firefighter Monica's residence. Firefighter Monica responded to the fire station and then returned to his home with other firefighters. When they arrived at the scene, Firefighter Monica collapsed of a heart attack.

April 16, 2005—1030hrs
Sally Renée Clark, Firefighter
Age 49, Volunteer
Pleasant Ridge Volunteer Fire Department, Mississippi

Firefighter Clark was participating in pump operations training at a fire department fundraising event. She suddenly collapsed. Firefighters and EMS workers responded immediately and came to her aid.

An AED was applied and CPR was initiated. When paramedics arrived, she was transported to the hospital. She died as a result of the heart attack on April 18, 2005.

April 16, 2005—1536hrs
Justin Paul Faur, Firefighter
Age 23, Volunteer
Andover Volunteer Fire Department, Iowa

Firefighter Faur was employed at a farm where he found an unconscious co-worker in a manure pit. Firefighter Faur initiated the 9-1-1 system and then entered the manure pit to attempt a rescue. Firefighter Faur was overcome by the atmosphere in the manure pit and collapsed before he could effect a rescue.

Responding firefighters from Firefighter Faur's department arrived on the scene. They donned SCBA's and entered the pit to remove both men. Firefighter Faur was transported to the hospital, where he died on April 30, 2005.

The cause of death was listed as brain injuries due to exposure to manure gases.

April 17, 2005—0907hrs
Alfred A. Wohrman, Firefighter
Age 60, Volunteer
Beekman Fire District, New York

Firefighter Wohrman was driving a fire department rescue truck to the scene of a motor vehicle crash. The response of his apparatus was cancelled when it was determined that the incident lay outside his department's jurisdiction.

He was returning to the station when, for some unknown reason, the apparatus left the roadway and crashed into some trees. No brake or skid marks were found. It is not known whether Firefighter Wohrman was wearing his seatbelt.

Responding firefighters had to extricate Firefighter Wohrman from the apparatus. He was transported to the hospital, where he later died as a result of his injuries. An autopsy was unable to determine whether an underlying medical condition had contributed to the crash. The official cause of death was listed as blunt force trauma.

April 18, 2005—1522hrs
Jacob Earl Cook, Firefighter
Age 23, Volunteer
Evanston Fire Department, Wyoming

Robert Perjue Henderson, Assistant Lieutenant
Age 38, Volunteer
Evanston Fire Department, Wyoming

Members of the Evanston Fire Department were dispatched to a report of a structure fire in a three-level, end-unit townhouse. Lieutenant Henderson and Firefighter Cook were occupants of Engine 1, the first fire apparatus to arrive. Upon their arrival they reported smoke showing. Neighbors advised the firefighters that children were trapped on the second floor of the residence.

A handline was pulled by another firefighter; Lieutenant Henderson and Firefighter Cook advanced the line into the structure after donning protective clothing and SCBA. The firefighters advanced the hoseline to the second floor of the structure. Lieutenant Henderson appeared at the door and requested a thermal imaging camera. The camera was provided and Lieutenant Henderson returned to the second floor.

continued on next page

A backup crew with a handline was entering the front door of the structure when an explosion occurred. The explosion blew the backup firefighters away from the building and caused windows to be blown out. After the explosion, the fire progressed rapidly. The backup crew was unable to gain access to the second floor. Another handline advanced from the rear entrance was able to knock down the fire sufficiently to allow access to the second floor. Approximately 15 minutes after the explosion, firefighters were able to remove Lieutenant Henderson and Firefighter Cook from the structure. Both men were dead.

The cause of death for both firefighters was listed as smoke inhalation and burns over 50 percent of their bodies.

The fire was burning in a concealed space. When a door was opened on the second floor, the fire extended with explosive force. The fire was electrical in nature, caused when storage boxes and floor planking were laid on top of electrical wires.

For additional information regarding this incident, please refer to NIOSH Fire Fighter Fatality Investigation and Prevention Program report F2005-13 (http://www.cdc.gov/niosh/fire/reports/face200513.html).

April 20, 2005—1101hrs
David Wayne O'Conner, Driver/Operator
Age 38, Career
Memphis Fire Department, Tennessee

Driver/Operator O'Conner and his engine crew responded to a fire alarm in a building. The alarm pull was determined to be false, and the engine was headed back to quarters.

Driver/Operator O'Conner lost consciousness as the engine proceeded down the street. The engine's company officer was able to reach over the engine enclosure and activate the vehicle's parking brake. The apparatus came to a stop on the sidewalk.

Firefighters immediately removed Driver/Operator O'Conner from the driver's seat and initiated treatment. An ambulance was summoned. Driver/Operator O'Conner was transported to the hospital, where he was pronounced dead as a result of a CVA.

For additional information regarding this incident, please refer to NIOSH Fire Fighter Fatality Investigation and Prevention Program report F2005-17 (http://www.cdc.gov/niosh/fire/reports/face200517.html).

April 20, 2005—1850hrs
Brian Bruns, Pilot
Age 47, Wildland Contract
Aero Union Corporation, Chico, California

Paul Cockrell, Pilot
Age 52, Wildland Contract
Aero Union Corporation, Chico, California

Tom Lynch, Pilot
Age 41, Wildland Contract
Aero Union Corporation, Chico, California

Pilots Bruns, Cockrell, and Lynch were practicing water drops and undergoing training/qualification flight checks on a Lockheed P-3B Orion airtanker. The pilots were on their 10th training run of the day, having just refilled the aircraft's water tanks.

For an unknown reason, the aircraft crashed about 12 miles north of the Chico Municipal Airport. No distress call was received from the aircraft. All three pilots were killed in the crash or in the ensuing fire.

For additional information about this crash, consult the National Transportation Safety Board Web site at http://www.ntsb.gov/ntsb/query.asp—NTSB identification SEA05MA085.

April 25, 2005—2204hrs
Christopher Brian Hunton, Firefighter
Age 27, Career
Amarillo Fire Department, Texas

The Amarillo Fire Department received a report of a house on fire at 2200hrs on March 23, 2005. Ladder 1, an American LaFrance quint, departed Amarillo Station 1 at 2201hrs en route to the fire. As Ladder 1 turned off South Van Buren Street onto East 3rd Avenue, the left rear passenger door on the apparatus opened.

Firefighter Hunton, who was preparing to don his breathing apparatus and who was not wearing a safety belt, fell from the apparatus, striking his head on the street and sustaining severe head injuries. Firefighter Hunton was transported to a local hospital, where he underwent emergency surgery.

continued on next page

Firefighter Hunton's condition deteriorated, and he died of his injuries at 0953hrs on April 25, 2005.

The latching mechanism for the door through which Firefighter Hunton fell was found to be malfunctioning.

The summary above is taken from a thorough report on this incident that was prepared by the Texas State Fire Marshal. The report is available at http://www. tdi.state.tx.us/fire/fmloddinvesti.html

May 7, 2005—1226hrs
Michael Thomas Childress, Assistant Fire Chief
Age 48, Career
Level Cross Volunteer Fire Department, North Carolina

Assistant Chief Childress worked a 24-hour shift on May 6, 2005. During the shift, he performed normal fire station duties and responded to an EMS incident. In the morning, he was supposed to go to a nearby town for training. He decided to remain at the fire station to provide coverage, and because he was not feeling well. Prior to the departure of other firefighters, Assistant Chief Childress helped with the loading of 5-inch hose into a pickup truck for the training session.

About 1000hrs, Assistant Chief Childress told the other firefighter on duty that he was going to lie down. He proceeded to the dorm area and went to bed. About 1215hrs, Assistant Chief Childress's daughter stopped by the fire station and went to check on her father. She found him unresponsive. The other firefighter on duty checked and found that Chief Childress was pulseless and cool to the touch.

An autopsy revealed that Assistant Chief Childress had suffered from severe coronary artery disease.

May 14, 2005—2335hrs
Mark Mianulli, Firefighter
Age 51, Volunteer
Syosset Volunteer Fire Department, New York

The Syosset Fire Department was dispatched at 1649hrs for a vehicle crash involving an overturned fuel truck. Firefighter Mianulli responded to the fire station, but arrived too late to board the engine that responded from the station.

Firefighter Mianulli stood by at the station in case any other assistance was needed.

The incident involved a major spill of fuel and required mutual aid to control. Later in the evening, Firefighter Mianulli helped to load a large number of 50-pound bags of speed dry into a pickup truck at the fire station. After the pickup departed for the incident scene, Firefighter Mianulli remained on standby.

The engine returned to the fire station at 2035hrs. Firefighter Mianulli left the fire station between 2100hrs and 2130hrs.

At 2211hrs, firefighters from a nearby fire district were dispatched to a videotape rental store in their district; Firefighter Mianulli had suffered a heart attack while in the store. Firefighter Mianulli was transported to a local hospital, where he was pronounced dead.

May 19, 2005—1700hrs
Karl Klifton "Kliff" Kramer IV, Firefighter Recruit
Age 22, Career
Jacksonville Fire-Rescue, Florida

Firefighter Recruit Kramer and his class spent the morning and early afternoon engaged in high-angle-rescue training exercises. At the conclusion of the high-angle training, the class began a 3-mile run for physical training. The temperature was reported to be 79 degrees.

Approximately 2.5 miles into the run, Firefighter Recruit Kramer collapsed. He was placed into a vehicle and driven by other firefighters back to the training facility. An ambulance was called, and Firefighter Recruit Kramer was taken to the hospital.

Firefighter Recruit Kramer was found to be suffering from heat stroke, with a temperature of 108 degrees. He was rapidly cooled and provided with other advanced care. Despite aggressive treatment in the hospital, Firefighter Recruit Kramer died on May 28, 2005.

May 27, 2005—1337hrs
Frank Joseph Kucera, Lieutenant
Age 43, Career
Seminole County Fire Rescue, Florida

continued on next page

Lieutenant Kucera and the members of his crew participated in physically demanding SCBA air consumption training. During the training, Lieutenant Kucera told other firefighters that he was not feeling well. As the crew left the training area, Lieutenant Kucera directed his crew to drop him at a nearby fire station while the crew went for lunch.

After being dropped at the fire station, Lieutenant Kucera entered the fire station and called for help. Onduty firefighters brought Lieutenant Kucera to the day room area and an ambulance was called. Before the ambulance arrived, Lieutenant Kucera received ALS-level care. He was loaded into the arriving ambulance and transported to the hospital.

As the ambulance arrived at the hospital, Lieutenant Kucera went into full cardiac arrest. The time elapsed from the request for the ambulance to arrival at the hospital was 15 minutes. Despite 32 minutes of life-saving efforts in the emergency room, Lieutenant Kucera was not revived.

Lieutenant Kucera was a survivor of an aggressive cancer. The cause of death was cited as atherosclerotic cardiovascular disease with calcification-caused narrowing of the coronary arteries.

May 30, 2005—2205hrs
Robert Wayne Duff, Firefighter
Age 39, Volunteer
Kuttawa Fire Department, Kentucky

On May 30, 2005, at approximately 2205hrs, Firefighter Duff and his fire department were dispatched to a vehicle fire. He responded in his personal vehicle from his home. After he had responded approximately 8 miles, dispatch cancelled the response. Firefighter Duff returned to his residence.

The next day, Firefighter Duff went to work as normal. After work, he went to his sister's house. When he arrived at the house, he told family members that he did not feel well. He asked his family members to call an ambulance. The ambulance arrived and EMS workers started a line and placed Firefighter Duff on a heart monitor. While being treated, Firefighter Duff became unconscious.

Firefighter Duff was loaded into the ambulance and transported to a hospital. While en route to the hospital, Firefighter Duff went into cardiac arrest. Upon his arrival at the hospital, emergency room staff continued rescue efforts. Despite their efforts, Firefighter Duff was pronounced dead at 1722hrs. The cause of death was listed as a heart attack.

May 31, 2005—1425hrs
Paul Albert Carr, Firefighter
Age 58, Career
Atlantic City Fire Department, New Jersey

Firefighter Carr was completing physical fitness training using a rowing machine while on duty in the fire station. At some point, he became unconscious. Another firefighter entered the room an undetermined amount of time after Firefighter Carr became unconscious. The firefighter noticed that Firefighter Carr was unconscious and called for help from other firefighters.

An AED was applied, CPR was started, and an ambulance was requested. Firefighters continued to treat Firefighter Carr until his care was taken over by the ambulance crew. Firefighter Carr was carried by firefighters to the ambulance and then transported to the hospital. Firefighter Carr was pronounced dead at the hospital 34 minutes after he was found unconscious. His death was caused by atherosclerotic cardiovascular disease.

June 3, 2005—1500hrs
Audie Lee Cross, Fire Crew Supervisor
Age 44, Wildland Full-Time
Nevada Division of Forestry, Nevada

Fire Crew Supervisor Cross was managing a 12-person inmate conservation camp fire crew that was engaged in a conservation project. Fire Crew Supervisor Cross was operating an ATV. The ATV was involved in a rollover collision that resulted in the death of Fire Crew Supervisor Cross; the ATV was traveling downhill when it flipped over.

June 10, 2005—0044hrs
Bruce Sternberger, Firefighter
Age 56, Volunteer
Hardtner-Elwood Volunteer Fire Department, Kansas

Firefighter Sternberger and the members of his fire department fought a 260-acre fire in wheat fields. That night, a lightning strike occurred near his home, causing a power outage and starting a wildfire. Firefighter Sternberger called the fire department to report the fire and then went outside to begin fighting the fire.

continued on next page

Firefighters believe that Firefighter Sternberger left his home and almost immediately came into contact with an energized power line. After not hearing from him for a while, Firefighter Sternberger's wife called the fire department for assistance. Firefighter Sternberger's body was discovered a short time later.

June 10, 2005—Time Unknown
Terry A. Kelver, Lieutenant
Age 48, Volunteer
Wales Center Volunteer Fire Company, New York

Several hours after responding to numerous weather-related calls, Lieutenant Kelver was found collapsed in his place of employment.

June 13, 2005—0100hrs
James J. O'Neil, Firefighter
Age 54, Volunteer
Hempstead Fire Department, New York

Firefighter O'Neil responded to four fire alarms in the 24 hours prior to his death. The last incident was a car fire in a parking garage, reported at 0058hrs. At this incident, Firefighter O'Neil drove the pumper to the scene, assisted with stretching of an attack line through a window from the top of the apparatus, and then placed the apparatus back in service once the fire was controlled.

At the conclusion of the car fire incident, Firefighter O'Neill mentioned to other firefighters that he was experiencing chest discomfort, but dismissed it as insignificant.

At 0714hrs, an ambulance was requested to respond to Firefighter O'Neil's residence. Firefighter O'Neil had suffered a severe heart attack. CPR was begun and he was transported by ambulance to the hospital, where he was pronounced dead.

June 14, 2005—2300hrs
Peter Bruce Lund, Lieutenant
Age 54, Volunteer
Woodmere Fire Department, New York

Lieutenant Lund responded as the officer on his fire department's heavy rescue to a working fire in a residence. Lieutenant Lund participated in a search of the structure and opened up to allow engine company firefighters access to the fire in the attic. After the fire suppression operations were completed, Lieutenant Lund left the structure where the fire had occurred and sat down on a curb to rest. A firefighter walking by Lieutenant Lund asked him if he was feeling well. Seconds later, Lieutenant Lund lost consciousness and was found to be pulseless and not breathing.

Emergency medical aid was summoned, CPR was immediately started, and an AED was applied as ALS equipment was readied. These interventions were not successful, and Lieutenant Lund was loaded into an ambulance and transported to a hospital. Despite these efforts, Lieutenant Lund was pronounced dead at the hospital. The cause of death was listed as a heart attack.

Lieutenant Lund had recently retired from a distinguished career as a rescue company officer with the Fire Department of New York.

June 21, 2005—1330hrs
Keith David Allred, Fire Chief
Age 52, Volunteer
Juab Special Service Fire District—Granite Station of the West Desert Fire Department, Utah

Chief Allred was driving a fire department tanker (tender) to a safety inspection. The apparatus was a military surplus Army deuce-and-a-half. While he was driving on a dirt road at a speed estimated at 40 miles per hour, the left front tire blew out. The vehicle crossed over to the left side of the roadway, entered a ditch, and rolled over multiple times. Chief Allred was ejected in the course of the crash; he was apparently not wearing his seatbelt at the time of the crash.

A passerby came upon the crash scene and summoned help. Chief Allred was dead when law enforcement and EMS responders arrived.

June 21, 2005—1445hrs
Kenneth Ray Gailley, Firefighter/EMT
Age 45, Volunteer
Locust Community Volunteer Fire Department, Texas

continued on next page

Firefighter/EMT Gailley came upon a motor vehicle crash while driving a truck for his employer. He was the first emergency responder on the scene, and provided aid and treatment for the drivers involved in the crash. Upon the arrival of the Sherman Fire Department and the completion of the incident, Command released Firefighter/EMT Gailley to return to work.

The Locust Community Volunteer Fire Department and the Sherman Fire Department have a written mutual-aid agreement, and the Locust Community Volunteer Fire Department has a department policy for all EMT personnel to stop and render aid at any crash scene that they happen upon.

Ten minutes after departing the crash scene, Firefighter/EMT Gailley began to experience severe chest pains, and radioed ahead for EMS to respond to his place of employment. When he arrived at his place of employment about 2 minutes later, he went into cardiac arrest. He was treated by EMS personnel and transported to the hospital. He was pronounced dead at the hospital at 1725hrs.

June 22, 2005—1745hrs
William McAnally, Firefighter
Age 64, Volunteer
Ossining Fire Department, New York

Firefighter McAnally died suddenly while marching in a fire department parade. His death was caused by a heart attack.

June 23, 2005—1424hrs
Joseph William "Pete" Buckel, Fire Chief
Age 67, Volunteer
Bittinger Volunteer Fire Department, Maryland

Chief Buckel was responding from his residence to a 9-1-1 call. As he proceeded to his vehicle to begin his response, he suffered a heart attack and collapsed. His department was called to respond to his residence, but he could not be revived.

June 24, 2005—0245hrs
Valeree Sue Claude, Firefighter/EMT
Age 33, Career
Pinetop Fire Department, Arizona

Firefighter/EMT Claude responded to two incidents during the daylight portion of her shift. After dinner, she went to her bunkroom. At 0205hrs, her crew was dispatched to an incident but she did not respond. A crew member went to her bunkroom to check on her and found her to be unresponsive. EMS treatment was begun, and continued until her death was confirmed by paramedics. According to the medical examiner, Firefighter/EMT Claude died of an overdose of oxycodone, and other medical conditions.

For additional information regarding this incident, please refer to NIOSH Fire Fighter Fatality Investigation and Prevention Program report F2005-20 (http://www.cdc.gov/niosh/fire/reports/face200520.html).

June 28, 2005—1300hrs
John R. Husser, Ex-Captain
Age 55, Volunteer
Rockville Centre Fire Department, New York

Ex-Captain Husser responded to three emergency calls with his fire department on June 27, 2005. The calls included a structure fire response and two motor vehicle crashes. The last incident occurred at 1826hrs.

At 1300hrs on the following day, Ex-Captain Husser was found deceased in his residence, wearing the same clothes that he had worn to his emergency responses the previous day. The television in the room was tuned to the same channel that he had been seen watching at 2300hrs the previous evening. During the responses the previous day, Ex-Captain Husser had not expressed any health concerns to other firefighters. His death was caused by a heart attack.

June 28, 2005—2231hrs
Gary Wayne Jolley, Firefighter
Age 52, Volunteer
Mount Carmel Volunteer Fire Department, Kentucky

Firefighter Jolley and the members of his fire department responded to a report of a lightning strike on a residence. Firefighter Jolley responded directly to the scene in his personal vehicle. Firefighters thoroughly checked the home, including the attic, and found no evidence of fire. They advised the home owner to have the electrical system checked, and then left the scene to return to the fire station.

continued on next page

As Firefighter Jolley drove his personal vehicle to the fire station, he became ill and pulled his vehicle over to the side of the road. A report of a person slumped over the steering wheel of a car brought the response of firefighters, law enforcement, and ambulance personnel.

Firefighters pulled Firefighter Jolley from his vehicle and initiated CPR. An AED was applied and arriving paramedics took over care. Firefighter Jolley was transported to the hospital, where he was pronounced dead.

July 7, 2005—0135hrs
Thomas Allen Hurlbert, Jr., Fire Apparatus Operator
Age 54, Career
Rockdale County Fire Department, Georgia

Fire Apparatus Operator Hurlbert and the members of his engine company responded to a report of a tree fallen in the road as the result of storm activity. The shift had been especially busy, with 12 responses by Hurlbert's engine company. The crew removed the tree from the road. Fire Apparatus Operator Hurlbert returned to the engine and removed his protective clothing.

As Fire Apparatus Operator Hurlbert was pulling the engine away from the scene, his company officer heard him make an unusual noise. The company officer looked over and saw Fire Apparatus Operator Hurlbert lying doubled over in his seat. The officer was able to bring the apparatus to a stop by pulling the air brake.

Firefighters removed Fire Apparatus Operator Hurlbert from the vehicle. He stopped breathing, and no pulse could be located. Firefighters started CPR and applied an AED. He was transported to the hospital, where he died.

The cause of death was listed as a heart attack.

July 10, 2005—0130hrs
Joseph Harold Evans, Firefighter
Age 61, Volunteer
Bridgeville Volunteer Fire Company, Delaware

Firefighter Evans and members of his fire department responded to a report of a fire alarm activation. Firefighter Evans was the driver of an engine company that responded to the incident.

After his arrival on the scene, Firefighter Evans collapsed from a heart attack. Other firefighters came to his aid immediately, and he was transported to a hospital, where Firefighter Evans was pronounced dead.

July 14, 2005—0042hrs
Todd Alan Blanchard, Firefighter
Age 31, Career
Eastern Wake Fire-Rescue, North Carolina

Firefighter Blanchard and his engine company responded to a report of a brush fire. Firefighters arrived on the scene and found a fire in a large hollow oak tree, which had been the subject of a previous response. The Fire Chief sent a second engine with a large water tank to the scene to assist. When the second engine arrived, the crew used the deck gun on the pumper to apply water from a distance. They used the 1,000 gallons of water that was onboard the pumper, but failed to extinguish the fire completely.

Four firefighters approached the tree to assess the impact of the water application. Two of the firefighters returned to the apparatus and two firefighters, including Firefighter Blanchard, remained near the tree. Firefighters near the apparatus heard cracking sounds and turned to see a large tree limb falling from the tree.

Firefighter Blanchard was struck by the tree limb, whose diameter was reported to be 24 inches. Firefighters rushed to his location and got him out from under the limb. Medical treatment was initiated and an ambulance was requested. Firefighter Blanchard was transported to the hospital, but did not survive his injuries.

July 17, 2005—1700hrs
Donald Eugene "Dan" DeVries, Firefighter
Age 59, Volunteer
Belvidere Fire Department, South Dakota

A fire began in a hay bailer located on Firefighter DeVries' property. Firefighter DeVries reported the fire to the fire department, and began to fight a brush fire that extended from the fire in the hay bailer.

continued on next page

When the fire department arrived, the brush fire was mostly extinguished, with the only remaining fire being that in the bailer. The fire in the bailer was extinguished, and firefighters who had responded to the incident began to talk with Firefighter DeVries about the incident. He suddenly collapsed.

Firefighters started CPR and an ambulance was requested. Firefighter DeVries did not survive the heart attack; he was pronounced dead at the hospital.

July 18, 2005—Time Unknown
Gerald Michael Martinez, Forestry Technician/SEAT Manager
Age 53, Wildland Full-Time
USDA Forest Service—Custer National Forest, South Dakota

Forestry Technician Martinez was on assignment to a fire in Cortez, Colorado, as a Single Engine Air Tanker (SEAT) Manager. He had been assigned to these duties since July 7, 2005. He performed his duties from 0830hrs through 2100hrs on July 17, 2005.

Forestry Technician Martinez was discovered dead by hotel room cleaners on the morning of July 18, 2005; he had died overnight. The cause of death was listed as a heart attack.

August 2, 2005—1800hrs
William Bostian, Fire Police Lieutenant
Age 62, Volunteer
West Webster Fire Department, New York

Fire Police Lieutenant Bostian had responded to four emergency calls throughout the day. After the conclusion of the last incident, Fire Police Lieutenant Bostian returned to the fire station to prepare for a fire department meeting that night (Fire Police Lieutenant Bostian was the acting president of his fire company).

At approximately 1800hrs, Fire Police Lieutenant Bostian asked a fire department paramedic to check his blood pressure. The paramedic noted that Fire Police Lieutenant Bostian's blood pressure was high and that his pulse was erratic; the paramedic arranged for Fire Police Lieutenant Bostian to be transported to the hospital.

At approximately 2115hrs, Fire Police Lieutenant Bostian went into cardiac arrest at the hospital. He was transferred to the Cardiac Intensive Care Unit, and a pacemaker was implanted on August 5.

Fire Police Lieutenant Bostian was scheduled to be discharged from the hospital, but he suffered another heart attack and died while in the hospital.

August 6, 2005—1539hrs
Christopher Matthew Kanton, Firefighter
Age 23, Career
California Department of Forestry/Riverside County Fire Department, California

Firefighter Kanton and his engine company were responding to a storm-related flooding emergency. Firefighter Kanton was riding in the rear jumpseat area of the engine.

The engine lost traction on a major freeway and left the roadway. The apparatus crossed several lanes of traffic, spun approximately 180 degrees, slid approximately 40 feet down an embankment, and hit trees. Firefighter Kanton was ejected and received major traumatic injuries when the apparatus rolled over him. The driver of the engine was also ejected. The crash occurred during severe weather and downpours of rain.

Firefighter Kanton and the driver of the apparatus were not wearing their seatbelts. The firefighter riding in the right front seat was wearing his seatbelt. The driver and the firefighter who was not ejected received minor to moderate injuries.

August 11, 2005—2105hrs
Edwin H. Berg, Fire Chief
Age 68, Volunteer
North Prairie Fire Department, Wisconsin

Chief Berg was attending a meeting of the village board representing the fire department. Chief Berg experienced a coughing attack, spat up blood, and collapsed.

Chief Berg was fighting lung and liver cancer. His death was caused by a pulmonary embolism.

August 14, 2005—Time Unknown
Rodney N. English, Sergeant
Age 48, Career
Detroit Fire Department, Michigan

continued on next page

Sergeant English became ill while on duty in the fire station, performing administrative duties. He was treated by fellow firefighters and transported to the hospital. His death was caused by a heart attack.

August 16, 2005—1115hrs
Nicholas Lee Yuselew, Captain
Age 47, Paid-on-Call
Pueblo of Zuni Fire/EMS Department, New Mexico

Captain Yuselew returned from a physical fitness activity and suffered a cardiac arrest.

August 17, 2005—2047hrs
Joseph F. Walsh, Fire Police Officer
Age 76, Volunteer
Keansburg Fire Department, New Jersey

Fire Police Officer Walsh was directing traffic at a minor hazardous materials incident at the Keansburg High School. He was standing at the entrance and assisting drivers in getting cars out of the parking lot. As Fire Police Officer Walsh performed his duties, he was struck by a car driven by an intoxicated driver. He received serious head and leg injuries. (The car left the scene after hitting Fire Police Officer Walsh.)

Fire Police Officer Walsh was wearing a reflective vest and was using a flashlight at the time he was struck.

Fire Police Officer Walsh was treated at the scene and flown by medical helicopter to a regional hospital. His care continued at the hospital, but he was pronounced dead at 0205hrs on August 18, 2005.

A 26-year-old woman was charged with driving while intoxicated, striking Fire Police Officer Walsh, and other charges. The blood-alcohol level of the intoxicated driver when she was tested was three times the legal limit. The woman eventually pleaded guilty to charges and accepted a 12-year sentence in State prison.

August 18, 2005—0551hrs
David W. Stautamoyer, Assistant Fire Chief
Age 57, Volunteer
Modoc Volunteer Fire Department, Indiana

Assistant Chief Stautamoyer and members of his department participated in a work detail at the fire station on the evening of August 17, 2005. Approximately seven firefighters and officers moved equipment from a tanker (tender) that was going out of service for the addition of a dump valve, and also reloaded hose onto pumpers that had been used at a structure fire earlier. The work detail lasted approximately 1.75 hours and was completed at approximately 2015hrs.

The next morning, Assistant Chief Stautamoyer collapsed in his home as the result of a heart attack. Firefighters and EMS responders responded and treated him in his home, and in the ambulance during transport. The ambulance was met by paramedics while en route to the hospital. Assistant Chief Stautamoyer was pronounced dead in the hospital emergency room.

August 25, 2005—0600hrs
Daniel Raymond Angert, Fire Police Officer
Age 44, Volunteer
Petrolia Volunteer Fire Company, Pennsylvania

On August 24, 2005, Fire Police Officer Angert and the members of his fire department responded to a structural fire. Fire Police Officer Angert blocked an approach to the fire scene and controlled traffic. The traffic pattern in front of the fire-involved structure was one-way in order to facilitate a tanker shuttle.

At approximately 1300hrs, Fire Police Officer Angert notified the fire police commander that he had been run over by a car. Fire Police Officer Angert had momentarily turned away from traffic. When he looked back, a car was upon him, and it ran over Fire Police Officer Angert's foot and ankle. He was also suffering from chest pains.

Fire Police Officer Angert was transported to the hospital to have his injury evaluated. He was released from the hospital later in the day, and told to stay off his foot and to see his personal physician.

The following morning, an ambulance was called to Fire Police Officer Angert's home. He had suffered a heart attack. Bystander CPR was in progress when responders arrived. Fire Police Officer Angert was transported to the hospital, but did not survive.

August 27, 2005—2145hrs
Michael A. Switala, Firefighter
Age 50, Volunteer
Lower Burrell Volunteer Fire Department, Pennsylvania

continued on next page

Firefighter Switala was enrolled in a dive rescue training class offered by a local SCUBA enterprise. The class had two sessions, one in daylight and the other at night. Firefighter Switala successfully completed the daylight portion and had completed all of the structured exercises for the night portion of the class.

At the completion of the dive exercise, Firefighter Switala's dive partner signaled that he was ready to surface. Firefighter Switala acknowledged the signal and the dive partner surfaced. Once on the surface, the dive partner could not locate Firefighter Switala. The partner looked back under water and saw that Firefighter Switala was in distress. The partner descended to Firefighter Switala and attempted to place Firefighter Switala's regulator back into his mouth, to no avail. The partner attempted to place an emergency air supply in Firefighter Switala's mouth, again to no avail.

The partner then brought Firefighter Switala to the surface, called for help, and brought Firefighter Switala to shore. Once he was on shore, medical care was provided by a physician, a registered nurse, and an EMT. Firefighter Switala was transported by ambulance to a local hospital, and was then transferred to a regional hospital. Firefighter Switala died on August 28, 2005.

An examination of Firefighter Switala's dive gear found that he had not exhausted his air supply. The cause of death was listed as drowning.

September 2, 2005—2300hrs
Robert Allen Bestgen, Firefighter
Age 51, Volunteer
Osborn Fire Protection District, Missouri

Osborn firefighters routinely provided standby coverage at each race held at the U.S. 36 Raceway in Osborn. Firefighter Bestgen was participating in a standby on September 2, 2005.

During a race, a crash occurred on the track. Race cars blocked the entrance of a push vehicle onto the track. The driver of the push vehicle parked the truck and asked a car driver to move his vehicle to provide access to the track. The driver, not realizing the location of the push truck, backed up quickly and hit the front of the push truck. The collision pushed the truck backwards toward a crowd of spectators.

Firefighter Bestgen, seeing the movement of the truck and the proximity of spectators, attempted to enter the push truck and bring it to a stop. The left front wheel of the truck ran over Firefighter Bestgen's foot and dragged him under the truck. The truck crashed into a concrete barrier, went airborne, and landed on top of Firefighter Bestgen.

A tow truck was used to lift the vehicle off Firefighter Bestgen. He was transported by ambulance to a local hospital, where he died of chest and other internal injuries at 0134hrs on September 3, 2005.

September 4, 2005—1713hrs
Henry James Combs, Fire Chief
Age 46, Volunteer
Watts Volunteer Fire Department, Kentucky

Chief Combs and the members of his fire department were dispatched to a report of a structure fire in a residence. Chief Combs was responding with lights and siren to the scene in his personal vehicle, a 2002 Toyota Tacoma.

As the vehicle entered a left-hand turn, Chief Combs lost control and the vehicle started to go off the right-hand side of the road. Chief Combs steered to the left, causing the vehicle to skid out of control and overturn.

Local residents came upon the crash and found Chief Combs' vehicle on its roof. Thinking that Chief Combs was trapped under the vehicle, the residents pushed the vehicle back on its wheels. After the vehicle was turned over, the local residents discovered that Chief Combs was still in his vehicle, wearing his seatbelt. Chief Combs was transported to the hospital, where he was pronounced dead.

The cause of death was listed as positional asphyxiation, as a result of the upside-down position of his vehicle after the crash.

September 5, 2005—0130hrs
Marvin Jackson, Fire Chief
Age Unknown, Volunteer
Wilmar Fire Department, Arkansas

Chief Jackson and the members of his fire department responded to the scene of a vehicle fire. Upon their arrival, firefighters found a fully involved car. Chief Jackson operated the pump and other firefighters attacked the fire.

As the fire was being controlled, firefighters heard yelling and looked over to see civilians running toward the fire truck. Firefighters found Chief Jackson lying on the sidewalk. They found that he had shallow respirations and a weak pulse.

continued on next page

Shortly thereafter, Chief Jackson stopped breathing, and CPR was started. Chief Jackson was then transported to the hospital, where he died. The cause of death was a heart attack.

September 11, 2005—2006hrs
Robert Lee Pauley, Firefighter
Age 55, Volunteer
Lookout Volunteer Fire Department, California

Firefighter Pauley was responding in his personal vehicle to a motor vehicle crash. As he responded, he lost control of his vehicle and struck a telephone pole. Firefighter Pauley was ejected and thrown 50 feet from the vehicle. After the collision, the vehicle caught fire and became fully involved.

The crash was reported by a passing vehicle. Firefighters responded, extinguished the fire, and found Firefighter Pauley's body.

September 16, 2005—0530hrs
James Edward Scott, Assistant Chief
Age 58, Volunteer
Hamilton Volunteer Fire and EMS, Inc., North Carolina

On September 15, 2005, Assistant Chief Scott and members of his fire department responded to a report of a smell of smoke in a residence. Firefighters found no emergency, and all units went back into service. The incident was concluded at 1717hrs.

As Assistant Chief Scott was driving to work the next day, he complained to his passenger of having experienced chest pain the night before, and then slumped forward onto the steering wheel of his vehicle. The passenger was able to stop the vehicle safely. Assistant Chief Scott was found to be without respirations or a pulse.

EMS was called, and CPR was initiated. Assistant Chief Scott was transported to the hospital, where he was treated, without response. He was pronounced dead at 0705hrs.

September 21, 2005—1600hrs
Edwin E. King, Fire Chief
Age 55, Volunteer
Reno County Fire District Number 7, Kansas

Chief King was driving a brush truck in response to a controlled burn that had become uncontained. Conditions were smoky and the dirt road that Chief King was driving on was only 20 feet wide.

Chief King's brush truck collided head-on with his department's tanker (tender) as it responded to the same incident. Chief King was pronounced dead at the scene; the driver of the tanker was not injured.

September 22, 2005—2034hrs
Paul A. Belmarez, Captain
Age 45, Career
South Bend Fire Department, Indiana

Captain Belmarez worked his regular shift from 0700hrs on September 21, 2005, through 0700hrs on September 22, 2005. During the shift, Captain Belmarez and his crew worked to get a broken fire hydrant to close, and performed other routine station duties. Captain Belmarez also conducted training by donning and doffing SCBA's, and walked on the treadmill.

Captain Belmarez complained to other firefighters of not feeling well. He complained of his stomach being upset and did not eat.

At 2034hrs, South Bend Fire Department units were dispatched to Captain Belmarez's home for a report of a subject having difficulty breathing. Captain Belmarez was suffering from a heart attack. He was treated by paramedics at the scene and then transported to the hospital, where he died.

October 4, 2005—2209hrs
Peter E. Hotaling, Sr., Fire Police Captain
Age 57, Volunteer
Claverack Fire District—A.B. Shaw Fire Company, New York

Fire Police Captain Hotaling participated in his fire department's weekly maintenance and training detail. During the detail, he told the Fire Chief that he did not feel well and went home.

continued on next page

Three hours after leaving the detail, Fire Police Captain Hotaling suffered a fatal heart attack.

October 23, 2005—1230hrs
Paul Herbert Thorne, Sr., Firefighter
Age 71, Volunteer
Forestville Volunteer Fire Department, Maryland

Firefighter Thorne suffered from a CVA while helping to unload supplies at the fire station. He died 2 days later as a result of his illness.

October 23, 2005—1245hrs
Robert Gerald Gallardy, Captain
Age 47, Career
Altoona Fire Department, Pennsylvania

The Pennsylvania State Fire Academy was conducting a Field Instructor Development Class, which includes 2 days of live-fire training. The class was completing the final burn on Sunday, October 23, 2005.

Captain Gallardy, serving as an instructor, was assigned to start the basement fire. He had checked the fire, and then went to the first floor to see whether the students and their instructor were ready to enter the basement. The instructor stated that they were ready. Captain Gallardy re-entered the basement via the interior stairs to add two skids to the fire. The students, with their instructor, entered the basement. They moved to the doorway of the burn room, extinguished the fire, and then vented the smoke and heat out through the exterior stairway. Upon returning to the burn room, they found Captain Gallardy inside the burn room. They radioed a Mayday and called for an RIT to respond to the exterior stairs. The student instructors removed Captain Gallardy to the stairs, and the RIT then moved him to the street.

The ALS ambulance, assigned to the training grounds during the class, and a paramedic instructor met the RIT as they exited the building. Captain Gallardy was treated and transported to Lewistown Hospital. The paramedic on scene requested a helicopter to meet the ambulance at the hospital to provide transportation to a burn facility. Captain Gallardy was flown to Lehigh Valley Hospital burn unit. He had received burns on 85 percent of his body and to his respiratory system.

Captain Gallardy was pronounced dead on Tuesday October 25, 2005 at 1107hrs. The cause of his collapse is unknown; the autopsy did not reveal any underlying health conditions that might have led to a collapse.

October 24, 2005—2020hrs
Wendell Anthony Jeffery, Lieutenant
Age 47, Career
Memphis Fire Department, Tennessee

Lieutenant Jeffery was on duty in the fire station day room when he began to have a seizure. An ambulance was called, and Lieutenant Jeffery received treatment from other onduty firefighters.

In a short time, Lieutenant Jeffery went into cardiac arrest. He was transported to the hospital but was pronounced dead at 2120hrs, with the cause of death listed as a heart attack. Lieutenant Jeffery and his crew had responded to four calls during the shift before he became ill.

October 25, 2005—1425hrs
Franklin W. Eubanks, Firefighter
Age 47, Volunteer
Arlington Volunteer Fire Department, Mississippi

Firefighter Eubanks and the members of his department had just completed extinguishing a small wildland fire when Firefighter Eubanks suddenly collapsed from what was apparently a heart attack.

Firefighter Eubanks was transported by helicopter to a hospital, but he did not survive.

October 26, 2005—1030hrs
Walter Leo Sykes, Jr., Firefighter
Age 48, Volunteer
Lewiston Volunteer Fire Department, California

Firefighter Sykes and members of his fire department responded to a structure fire in a historic home in their community. Firefighter Sykes responded on a pumper with another firefighter.

continued on next page

When they arrived on the scene, Firefighter Sykes walked to the fire building and started to feel ill. He stood near the front fence and told another firefighter that he felt tired. Firefighter Sykes was directed to the back of a pumper to rest. A firefighter noticed that he did not look well, and placed him in a car to bring him to an ambulance on the scene.

Firefighter Sykes was placed on oxygen, but his condition deteriorated. The EMS crew called for a helicopter, but none was available because of bad weather. Firefighter Sykes was transported by ambulance to the hospital, where he died of a heart attack.

November 4, 2005—1019hrs
E. Timothy Parsell, Captain
Age 39, Volunteer
Collins Volunteer Fire Department #1, New York

Captain Parsell was at home when his fire department was dispatched to a medical emergency. As he walked through his home, preparing to leave for the dispatch, Captain Parsell became lightheaded and fell to the floor.

Captain Parsell's wife called 9-1-1 and reported the incident. Captain Parsell was suffering from chest pains and was having difficulty breathing. While being transported by ambulance to the hospital, Captain Parsell went into cardiac arrest. He was later pronounced dead at the hospital, with death caused by a pulmonary embolism.

November 5, 2005—1753hrs
Eduardo B. "Ed" Teran, Firefighter
Age 43, Career
City of Riverside Fire Department, California

Firefighter Teran and other firefighters were fighting a fire in a vacant residential structure. Firefighter Teran had left the interior of the structure to change his SCBA cylinder when he collapsed of an apparent heart attack.

Firefighters on the scene provided care, and Firefighter Teran was transported to the hospital, but he did not survive the heart attack. The fire was caused by arson.

November 5, 2005—1130hrs
James C. "Jimmy" Webb, Assistant Chief
Age 41, Volunteer
Bynum Volunteer Fire Department, Mississippi

Assistant Chief Webb was driving a fire department tanker (tender) to a structure fire. He lost control of the apparatus and it overturned in a ditch. Chief Webb was ejected from the vehicle during the crash and died of traumatic injuries.

November 6, 2005—2230hrs
Brian J. Pugh, Fire Chief
Age 53, Career
Port of Portland Airport Fire Rescue, Oregon

Chief Pugh was attending a fire service management conference in Dallas, Texas, when he suffered a CVA; he died 2 days later.

November 7, 2005—1332hrs
Timmy Shane Hardy, Engineer
Age 32, Career
Neosho Fire Department, Missouri

Engineer Hardy and members of his fire department were dispatched to a report of a fire in a mill. Upon the arrival of firefighters, mill workers explained that they were dealing with a fire in a large bin that contained wheat hulls. The wheat hull bin was one of five in a 100-foot tall silo; the only access to the top of the silo was by means of a man lift.

In order to assess the status of the fire, two firefighters were assigned to ride the man lift to the top of the silo. Mill employees provided firefighters with information on the operation of the man lift. Engineer Hardy was the first to ascend, wearing full structural firefighting protective clothing.

Near the top of the silo, Engineer Hardy became wedged in the confined space between the man lift and the structural members of the landing platform. Apparently, Engineer Hardy's SCBA caught on the structural member. The force of the man lift bent Engineer Hardy backwards and did not allow him to operate the man-lift controls. Engineer Hardy was obviously dead when he was reached by plant employees and firefighters.

continued on next page

Engineer Hardy's body was removed from the silo after the arrival of a ladder truck from the Joplin Fire Department. The cause of death was listed as positional asphyxia. The original fire was determined to be accidental, caused by sparks from welding.

November 10, 2005—1050hrs
Kenneth D. Mitchell, Fire Chief
Age 58, Volunteer
Tull Fire and Rescue, Arkansas

Chief Mitchell had just arrived on the scene of a working structure fire in a residence. He began to feel ill and went to sit in his vehicle. Other firefighters found him in distress. Firefighters summoned an ambulance, an AED was attached to Chief Mitchell, and paramedics provided treatment.

Chief Mitchell was pronounced dead of an apparent heart attack. The house involved with the fire was unoccupied and had been used to manufacture illegal drugs in the past. The cause of the fire was last listed as "under investigation."

November 12, 2005—1445hrs
Kevin Joe Foster, Captain
Age 41, Volunteer
Ellerslie Volunteer Fire Department, Georgia

Captain Foster was the driver and only occupant of a 1990 Emergency One pumper. He had just cleared the scene of an EMS incident and was dispatched to respond to a motor vehicle crash in a neighboring community.

As the apparatus responded at a speed estimated by witnesses as 35 miles per hour, the right wheels of the pumper left the pavement and went off the road to the right, where there was a drop-off of approximately 3 inches. Captain Foster steered to the left; the apparatus crossed over into the oncoming lane and then left the roadway on the left-hand side. After the pumper left the roadway, it became airborne over a culvert and traveled 53 feet until it hit an oak tree and came to rest.

Captain Foster, who was not wearing a seatbelt, was thrown onto the floor on the passenger side of the pumper. Civilians who came upon the crash removed Captain Foster from the apparatus and attempted to provide CPR. Their efforts were not successful, and Captain Foster was pronounced dead at the scene.

November 12, 2005—Time Unknown
Max Buford Willard, Fire Chief
Age 69, Volunteer
Oakwood Volunteer Fire Department, Virginia

Chief Willard and the members of his fire department and other fire departments were fighting a wildland fire in a wooded area. The fire, which was caused by sparks from a lawnmower, eventually spread to approximately 350 acres. A total of approximately 70 firefighters were working the incident.

About an hour into the fire fight, Chief Willard entered the woods with a rake. It was approximately 1400hrs and Chief Willard was not accompanied by anyone. When Chief Willard did not return from his task after it had grown dark, firefighters began to search for him. Chief Willard's body was located at approximately 0830hrs on the following morning. He had suffered extensive burns.

Although the details of Chief Willard's death cannot be known, an investigator speculated that Chief Willard was overtaken by the fire while raking a fire line. He did not have time to escape uphill.

November 14, 2005—2345hrs
James Elvis Lafferty, Sr., Fire Police Captain
Age 68, Volunteer
Union City Volunteer Fire Department, Pennsylvania

Fire Police Captain Lafferty and the members of his fire department responded to the scene of a fire in a two-story wood-frame residence. Shortly after firefighters arrived, the fully involved structure collapsed into the basement. Mutual aid from surrounding fire departments was requested to fight the fire.

Fire Police Captain Lafferty was working traffic control on the fire scene when he was struck with a heart attack. Firefighters immediately began CPR and were joined by paramedics who were on the scene of the fire. Fire Police Captain Lafferty was transported by ambulance to the hospital, where he was pronounced dead.

November 20, 2005—1430hrs
Carl Fredrick Lambert, Firefighter
Age 66, Volunteer
Cabo Lucero Volunteer Fire Department, New Mexico

Firefighter Lambert was attending a 3-day wildland-fire chainsaw training course offered by the New Mexico State Forestry Division. On the 3rd day of training, firefighters had felled two trees and were in the process of removing tree limbs when Firefighter Lambert expressed the need for a break.

After walking a short distance, Firefighter Lambert became unsteady on his feet. Firefighters came to his aid to give him support and he walked a short distance further, and then lost consciousness. EMS-trained personnel who were enrolled in the class, including EMT's and a paramedic, began to treat Firefighter Lambert.

Firefighter Lambert was transported by vehicle to meet an ambulance that had been requested. His treatment was transferred to ambulance staff, and he was transported to the hospital. Upon his arrival at the hospital, it was confirmed that Firefighter Lambert had suffered a heart attack. As he was being prepared for helicopter transfer to a regional hospital, he suffered a second heart attack.

Firefighter Lambert was stabilized and transferred to the regional hospital. He underwent emergency surgery, but did not survive.

November 22, 2005—1427hrs
Clint Dewayne Rice, Firefighter
Age 28, Volunteer
Carlton Volunteer Fire Department, Texas

Firefighter Rice was driving a tractor-trailer fire department tanker (tender) to a mutual-aid grass fire. Firefighter Rice lost control of the apparatus while rounding a curve; the apparatus left the roadway and overturned.

Firefighter Rice was ejected from the vehicle and was pronounced dead at the scene. He had been a member of his department for 3 months.

November 22, 2005—Time Unknown
Robert Timothy Staepel, Jr., Firefighter/EMT
Age 41, Career
Navy Region Mid-Atlantic, Philadelphia Federal Fire Department, Pennsylvania

Firefighter/EMT Staepel participated in live-fire training while on duty for his regular shift on November 21, 2005. The next day, he felt unwell and called 9-1-1 from his home. Shortly after placing the call, he went into cardiac arrest.

Firefighter/EMT Staepel never regained consciousness, and died on November 25, 2005.

November 23, 2005—2345hrs
Charles Arnold McKenzie, Firefighter
Age 75, Volunteer
West Van Lear Fire Department, Kentucky

Members of Firefighter McKenzie's fire department had been on the scene of a structural fire for approximately 2 hours. They had initially been pumping water from a nearby creek, but the creek ran dry. Firefighter McKenzie arrived at the incident and agreed to become a driver in a water tanker (tender) shuttle.

Firefighter McKenzie had delivered four loads of water and was waiting in line to dump his tank. For some reason, Firefighter McKenzie and his passenger got out of the truck and began walking toward the fire.

The passenger was walking in front of Firefighter McKenzie. He heard Firefighter McKenzie yell and turned to see Firefighter McKenzie pinned beneath his truck as it rolled toward a ditch. Firefighter McKenzie was crushed by the apparatus as it entered the ditch.

November 28, 2005—0939hrs
Christopher James Roy, Firefighter
Age 25, Career
Calera Fire/Rescue, Alabama

Firefighter Roy was responding as the driver and only occupant of a pumper en route to an EMS incident. A few minutes into the response, the pumper crashed into the trailer of a tractor trailor truck at an intersection. The collision caused major damage to the apparatus, with the damage focused on the driver's position.

Firefighters were dispatched to the scene after reports of the crash reached the dispatch center. Arriving firefighters found Firefighter Roy conscious, complaining that he could not feel his legs and that his chest felt funny. Firefighter Roy's legs were entangled in the wreckage and extrication was needed.

continued on next page

The extrication effort was significant, involving the use of a backhoe, a forklift, a tow truck, and hydraulic rescue tools. Firefighter Roy was extricated approximately 40 minutes after the crash. His condition had worsened during the extrication process. He was transported by ambulance to a nearby landing zone, and then transported by medical helicopter to a regional hospital, where Firefighter Roy died of his injuries.

A lawsuit was filed by Firefighter Roy's mother in January 2006, claiming that the drivers of the tractor-trailer truck and a dump truck failed to yield to an emergency vehicle.

December 2, 2005—1320hrs
Richard Bernard "Ricky" McCurley, Captain
Age 33, Career
New Orleans Fire Department, Louisiana

Captain McCurley's engine company was responding to a report of a gas leak. As the apparatus entered an intersection during the response, the driver swerved to avoid a tractor-trailer truck that also was entering the intersection. The pumper clipped a minivan stopped at the intersection, left the roadway, and rolled over.

The driver of the apparatus was ejected. Captain McCurley was pinned under the "A" pillar of the fire apparatus and was killed. The cause of death was listed as positional asphyxia. The speed of the fire apparatus was cited as a factor in the crash.

December 10, 2005—1948hrs
Chelsea Lyn Garvin, Firefighter
Age 19, Volunteer
Fish River-Marlow Fire and Rescue, Alabama

Firefighter Garvin and two other firefighters were returning from a standby at the Fish River Christmas Boat Parade. The firefighters were returning to their launch site in the department's 15-foot rescue boat about an hour after the conclusion of the parade.

The boat collided head-on with another vessel, and Firefighter Garvin and another firefighter were thrown into the water. Firefighter Garvin was pulled from the water by boaters and was found to be in cardiac arrest.

Firefighter Garvin was transported by helicopter to a regional hospital, where she died of her injuries on December 15, 2005.

December 11, 2005—0130hrs
Chad Ernest Wessels, Captain
Age 31, Volunteer
Briggs Volunteer Fire Department, Texas

Captain Wessels was the driver and only occupant of a 1,200-gallon fire department tanker (tender) responding to a structure fire. During the response, a rear wheel of the flatbed truck failed, causing the tanker to leave the roadway and crash.

The tanker struck a fence and some trees, and then caught fire. Captain Wessels was not wearing a seatbelt, and he was killed in the crash.

Captain Wessels was also a career firefighter at Fort Hood.

December 20, 2005—2138hrs
Michael A. Hart, Firefighter/Paramedic
Age 33, Career
Elkins Fire Department, West Virginia

Firefighter/Paramedic Hart was returning from teaching a Firefighter I class in his personal vehicle. A tractor-trailer swerved to avoid a stopped car and crashed into Firefighter/Paramedic Hart's vehicle. He was pronounced dead at the scene.

INCIDENTS PRIOR TO 2005

February 10, 1994
Dennis J. Bottge, Lieutenant
Age at Death 53, Career
Palm Beach County Fire-Rescue, Florida

Lieutenant Bottge was providing care to the victim of a cardiac arrest. The patient suffered a series of seizures, resulting in Lieutenant Bottge's significant exposure to blood and other body fluids. As a result of this exposure, Lieutenant Bottge contracted Hepatitis C.

He began treatment and continued to work until his heath deteriorated and he was no longer able to work. Lieutenant Bottge was granted disability benefits upon his retirement. He died of complications of his disease on May 22, 2005.

May 27, 2000
Donald G. Pedro, Lieutenant
Age at Death 45, Career
Endicott Fire Department, New York

Lieutenant Pedro was on duty at the fire station. As he was working on his personal vehicle, a jack buckled and the vehicle crushed him. Lieutenant Pedro was severely injured, and received a disability retirement.

Lieutenant Pedro suffered a brain injury, and as a result was unable to work or to live at home. He suffered a seizure on December 31, 2004, and died.

Unknown Date in 2000
William Hudson, Captain
Age at Death 50, Career
Salem Fire Department, Massachusetts

Captain Hudson contracted Hepatitis C while on duty in 2000, and died as the result of his disease on February 3, 2005.

June 26, 2004
Michael James Bevens, Captain
Age at Death 59, Career
North Little Rock Fire Department, Arkansas

Captain Bevens suffered a shoulder injury while loading an obese patient into an ambulance. He was assigned to the fire department fire marshal's office while on limited duty.

He underwent surgery for his injury on November 6, 2005, and died of complications of the surgery.

December 27, 2004—1032hrs
Thomas L. Ivey, Firefighter
Age at Death 48, Volunteer
West Iron County Fire Department, Michigan

Firefighter Ivey was responding to a motor vehicle crash in his personal vehicle when he was involved in a crash. He suffered a cervical fracture and spinal cord injuries that resulted in quadriplegia. He died of complications of the quadriplegia on July 26, 2005.

FIREFIGHTER FATALITY INCLUSION CRITERIA—NATIONAL FIRE SERVICE ORGANIZATIONS

The National Fire Protection Association (NFPA), the National Fallen Firefighters Foundation (NFFF), the U.S. Fire Administration (USFA), and other organizations collect information on firefighter fatalities in the United States. Each organization uses a slightly different set of inclusion criteria that are based at least in part on the purposes of the information collection for each organization and data consistency.

As a result of these differing inclusion criteria, statistics about firefighter fatalities may be provided by each organization that do not coincide with one another. This section will explain the inclusion criteria for each organization and provide information about these differences.

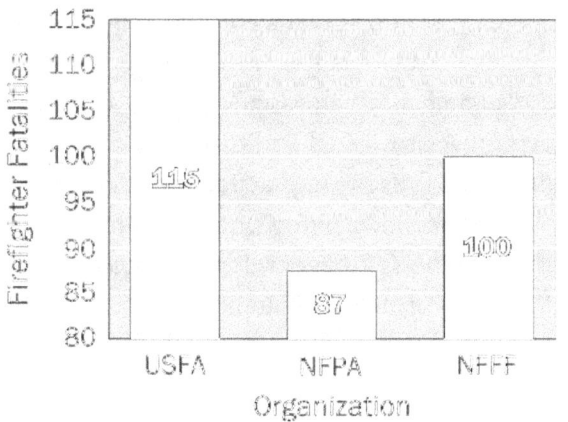

Firefighter Fatalities in 2005

The USFA includes firefighters in this report who die while on duty, become ill while on duty and later die, and firefighters who die within 24 hours of an emergency response or training regardless of whether the firefighter complained of illness while on duty. The USFA counts firefighter deaths that occur in the 50 States, the District of Columbia, and United States territories such as Puerto Rico and Guam.

For 2005, the USFA reported 115 onduty firefighter fatalities.

INCLUSION CRITERIA FOR NFPA'S ANNUAL FIREFIGHTER FATALITY STUDY

Introduction

Each year, the NFPA collects data on all firefighter fatalities in the United States that resulted from injuries or illnesses that occurred while the victims were on duty. The purpose of the study is to analyze trends in the types of illnesses and injuries resulting in death that occur while firefighters are on the job. This annual census of firefighter fatalities in its current format dates back to 1977. (Between 1974 and 1976, NFPA published a study of onduty firefighter fatalities that was not as comprehensive.)

What is a Firefighter?

For the purpose of the NFPA study, the term "firefighter" covers all uniformed members of organized fire departments, whether career, volunteer, combination, or contract; full-time public service officers acting as firefighters; State and Federal government fire service personnel; temporary fire suppression personnel operating under official auspices of one of the above; and privately employed firefighters including trained members of industrial or institutional fire brigades, whether full- or part-time.

Under this definition, the study includes, besides uniformed members of local career and volunteer fire departments, those seasonal and full-time employees of State and Federal agencies who have fire suppression responsibilities as part of their job description, prison inmates serving on firefighting crews, military personnel performing assigned fire suppression activities, civilian firefighters working at military installations, and members of industrial fire brigades. Impressed civilians also would be included if called on by the officer in charge of the incident to carry out specific duties. The NFPA study includes fatalities that occur in the 50 States and the District of Columbia.

What Does "On Duty" Mean?

The term "on duty" refers to being at the scene of an alarm, whether a fire or nonfire incident; being en route while responding to or returning from an alarm; performing other assigned duties such as training, maintenance, public education, inspection, investigations, court testimony, and fundraising; and being on call, under orders, or on standby duty other than at home or at the individual's place of business. Fatalities that occur at a firefighter's home may be counted if the actions of the firefighter at the time of injury involved firefighting or rescue.

Onduty fatalities include any injury sustained in the line of duty that proves fatal, any illness that was incurred as a result of actions while on duty that proves fatal, and fatal mishaps involving non-emergency occupational hazards that occur while on duty. The types of injuries included in the first category are mainly those that occur at an incident scene, in training, or in accidents while responding to or returning from alarms. Illnesses (including heart attacks) are included when the exposure or onset of symptoms are tied to a specific incident of onduty activity. Those symptoms must have been in evidence while the victim was on duty for the fatality to be included in the study.

Fatal injuries and illnesses are included even in cases where death is considerably delayed. When the onset of the condition and the death occur in different years, the incident is counted in the year of the condition's onset. Medical documentation specifically tying the death to the specific injury is required for inclusion of these cases in the study.

Categories Not Included in the Study

The NFPA study does not include members of fire department auxiliaries, nonuniformed employees of fire departments, emergency medical technicians (EMT's) who are not also firefighters, chaplains, or civilian dispatchers. The study also does not include suicides as onduty fatalities even when the suicide occurs on fire department property.

The NFPA recognizes that a comprehensive study of firefighter onduty fatalities would include chronic illnesses (such as cardiovascular disease and certain cancers) that prove fatal and that arose from occupational factors. In practice, there is as yet no mechanism for identifying onduty fatalities that are due to illnesses that develop over long periods of time. This creates an incomplete picture when comparing occupational illnesses to other factors as causes of firefighter deaths. This is recognized as a gap the size of which cannot be identified at this time because of the limitations in tracking the exposure of firefighters to toxic environments and substances and the potential long-term effects of such exposures.

2005 Experience

In 2005, a total of 87 onduty firefighter deaths occurred in the United States, according to NFPA's inclusion criteria.

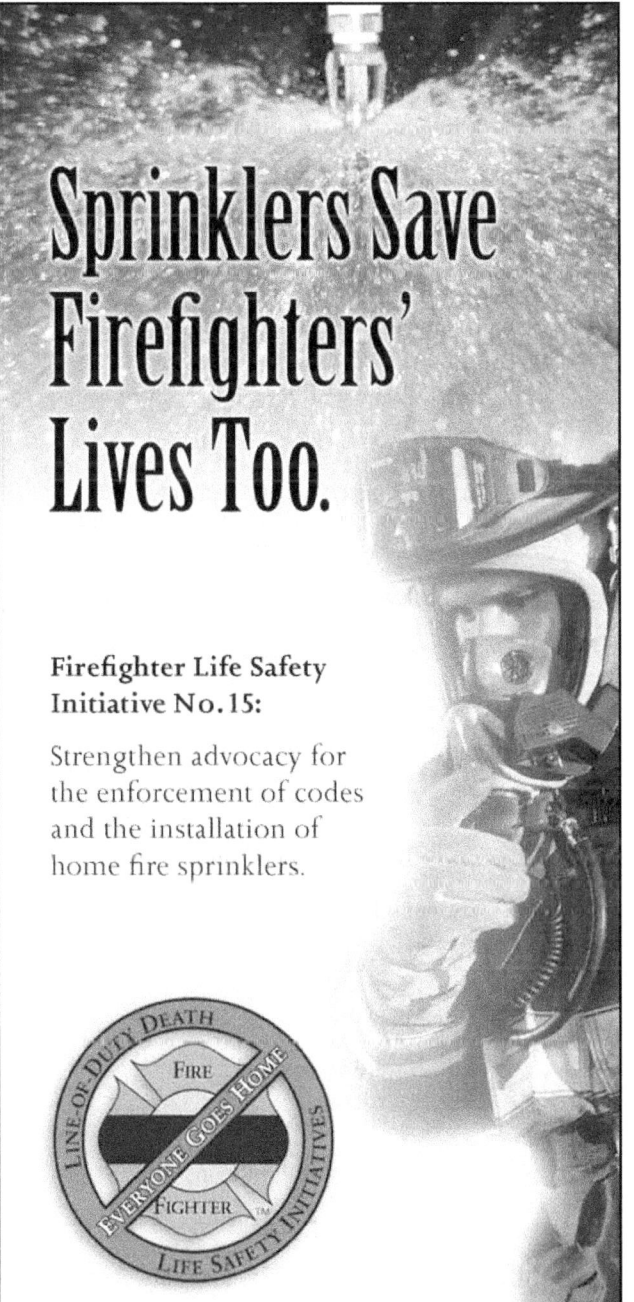

NATIONAL FALLEN FIREFIGHTERS FOUNDATION

In 1997, fire service leaders formulated new criteria to determine eligibility for inclusion on the National Fallen Firefighter Memorial. Line-of-duty deaths shall be determined by the following standards:

1. (a) Deaths of firefighters meeting the Department of Justice's Public Safety Officers' Benefits (PSOB) program guidelines, and those cases that appear to meet these guidelines whether or not PSOB staff has adjudicated the specific case prior to the annual National Fallen Firefighters Memorial Service; and

   (b) Deaths of firefighters from injuries, heart attacks, or illnesses documented to show a direct link to a specific emergency incident or department-mandated training activity.

2. While PSOB guidelines cover only public safety officers, the Foundation's criteria also include contract firefighters and firefighters employed by a private company, such as those in an industrial brigade, provided that the deaths meet the standards listed above.

3. Some specific cases will be excluded from consideration, such as deaths attributable to suicide, alcohol, or substance abuse, or other gross abuses as specified in the PSOB guidelines.

The National Fallen Firefighters Memorial was built in 1981 in Emmitsburg, Maryland. The names listed there begin with those firefighters who died in the line of duty that year. The United States Congress created the National Fallen Firefighters Foundation to lead a nationwide effort to remember America's fallen firefighters. Since 1992, the tax-exempt, nonprofit Foundation has developed and expanded programs to honor our fallen fire heroes and assist their families and coworkers by providing them with resources to rebuild their lives. Since 1997, the Foundation has managed the National Memorial Service held each October to honor the firefighters who died in the line of duty the previous year.

At the
October 2006 Memorial Weekend, the
Foundation will be honoring 107 firefighters who died in the line
of duty. Of those 107 being honored, 100 died in 2005 and 7 others died in
previous years. The following is a list of the 100 personnel to be honored
who died in 2005:

Keith David Allred, *Fire Chief*

Daniel Raymond Angert, *Fire Police Officer*

Michael Alfred Aunkst, *Firefighter*

John Gerard Bellew, *Lieutenant*

Robert Allen Bestgen, *Firefighter*

Michael James Bevens, *Captain**

Todd Alan Blanchard, *Firefighter*

William Bostian, *Fire Police Lieutenant*

Dennis J. Bottge, *Lieutenant**

Robert G. Brooks, Sr., *Firefighter Trainee*

Brian Bruns, *Pilot*

Joseph William "Pete" Buckel, *Fire Chief*

Gerald Joseph "Jerry" Buehne, *Fire Chief*

Grady Don Burke, *Captain*

Paul Albert Carr, *Firefighter*

Michael Thomas Childress, *Assistant Fire Chief*

Sally Renée Clark, *Firefighter*

Valeree Sue Claude, *Firefighter/EMT*

Paul Cockrell, *Pilot*

Henry James Combs, *Fire Chief*

Donald Conner, *Firefighter*

Jacob Earl Cook, *Firefighter*

Donald Eugene "Dan" DeVries, *Firefighter*

Christopher R. DeWolf, *Lieutenant*

Robert Wayne Duff, *Firefighter*

Charles Lynn Edgar, *Fire Management Officer*

Andre Miguel "Mike" Ellis, *Sergeant*

Franklin W. Eubanks, *Firefighter*

Joseph Harold Evans, *Firefighter*

Michael D. Falkouski, *Assistant Fire Chief*

Richard Allen Fast, *Firefighter*

Justin Paul Faur, *Firefighter*

Kevin Joe Foster, *Captain*

James H. Fugate, *Firefighter**

Ornell Edgar Fuller, Jr., *Captain*

Kenneth Ray Gailley, *Firefighter/EMT*

Robert Gerald Gallardy, *Captain*

José Victor Gonzales, *Pilot*

William "Bill" Matt Goodin, *Captain*

John Greeno, *Heliport Base Manager*

Timmy Shane Hardy, *Engineer*

Michael A. Hart, *Firefighter/Paramedic*

Robert Purjue Henderson, *Assistant Lieutenant*

Henry DeAngelo Hobbs, Jr., *Senior Forest Ranger*

Jerry Wayne Hopper, *Forestry Technician*

Christopher Brian Hunton, *Firefighter*

Thomas Allen Hurlbert, Jr., *Fire Apparatus Operator*

John R. Husser, *Firefighter*

Thomas L. Ivey, *Firefighter**

Marvin Jackson, *Fire Chief*

Wendell Anthony Jeffery, *Lieutenant*

Gary Wayne Jolley, *Firefighter*

Christopher Matthew Kanton, *Firefighter*

Terry A. Kelver, *Lieutenant*

Edwin E. King, *Fire Chief*

Karl Klifton "Kliff" Kramer IV, *Firefighter Recruit*

continued on next page

Frank Joseph Kucera, *Lieutenant*

James Elvis Lafferty, Sr., *Fire Police Captain*

Peter Bruce Lund, *Lieutenant*

Tom Lynch, *Pilot*

Gerald Michael Martinez, *SEAT Manager*

Mark Francis "Mac" McCormack, *Captain*

Richard Bernard "Ricky" McCurley, *Captain*

Charles Arnold McKenzie, *Firefighter*

Michael Angelo Mercurio, Jr., *Firefighter/EMT*

James E. Mero, Jr., *Deputy Fire Coordinator/Fire Inves.*

Curtis W. Meyran, *Lieutenant*

Mark A. Mianulli, *Firefighter*

Kenneth D. Mitchell, *Fire Chief*

Dale A. Monica, *Firefighter*

Thomas Logan Mower, *Fire Police Officer*

Lonnie Wayne Nicklas, *Fire Chief*

David Wayne O'Conner, *Driver/Operator*

James J. O'Neil, *Firefighter*

E. Timothy Parsell, *Captain*

Robert Lee Pauley, *Firefighter*

Angelo Petta, *Chief Engineer*

Brandon Scott Phillips, *Firefighter/Paramedic*

William Wade Pierce, *Firefighter*

William H. Poage, Jr., *Firefighter*

Clint Dewayne Rice, *Firefighter*

Christopher James Roy, *Firefighter*

Walter Matthew "Matt" Sarnoski, *Firefighter*

Richard Thomas Sclafani, *Firefighter*

Carl E. Sherman, *Firefighter*

Todd Raymond Smith, *Firefighter*

Robert Timothy Staepel, Jr., *Firefighter*

Bruce Sternberger, *Firefighter*

Michael A. Switala, *Firefighter*

Walter Leo Sykes, Jr., *Firefighter*

Eduardo B. "Ed" Teran, *Firefighter*

Scott Allen Thornton, *Captain*

Joseph F. Walsh, *Fire Police Officer*

James C. "Jimmy" Webb, *Assistant Chief*

Chad Ernest Wessels, *Captain*

Max Buford Willard, *Fire Chief*

Justin M. Wisniewski, *Firefighter*

Alfred A. Wohrman, *Firefighter*

Allen Wayne Wright, *Fire Chief*

Timmy Young, *Fire Equipment Operator*

The following firefighters who died prior to 2005 also are being honored:

Leroy M. "Punch" Byers, *Fire Police*

Eugene M. Knause, *Firefighter*

Vincent G. McGuinness, *Firefighter*

Bennie J. Shields, *Lieutenant*

Carl W. Shoemaker, *Engineer*

Gerald J. Winuk, *Firefighter/EMT*

Robert E. Woolf, *Firefighter*

---

* *Denotes a firefighter fatality that occurred in 2005 as the result of an incident in a previous year.*